Studies in Systems, Decision and Control

Volume 22

Series editor
Janusz Kacprzyk, Polish Academy of Sciences, Warsaw, Poland
e-mail: kacprzyk@ibspan.waw.pl

About this Series

The series "Studies in Systems, Decision and Control" (SSDC) covers both new developments and advances, as well as the state of the art, in the various areas of broadly perceived systems, decision making and control- quickly, up to date and with a high quality. The intent is to cover the theory, applications, and perspectives on the state of the art and future developments relevant to systems, decision making, control, complex processes and related areas, as embedded in the fields of engineering, computer science, physics, economics, social and life sciences, as well as the paradigms and methodologies behind them. The series contains monographs, textbooks, lecture notes and edited volumes in systems, decision making and control spanning the areas of Cyber-Physical Systems, Autonomous Systems, Sensor Networks, Control Systems, Energy Systems, Automotive Systems, Biological Systems, Vehicular Networking and Connected Vehicles, Aerospace Systems, Automation, Manufacturing, Smart Grids, Nonlinear Systems, Power Systems, Robotics, Social Systems, Economic Systems and other. Of particular value to both the contributors and the readership are the short publication timeframe and the world-wide distribution and exposure which enable both a wide and rapid dissemination of research output.

More information about this series at http://www.springer.com/series/13304

J. Christopher Westland

Structural Equation Models

From Paths to Networks

 Springer

J. Christopher Westland
Department of Information and Decision Sciences
University of Illinois
Chicago, IL, USA

ISSN 2198-4182 ISSN 2198-4190 (electronic)
Studies in Systems, Decision and Control
ISBN 978-3-319-38631-7 ISBN 978-3-319-16507-3 (eBook)
DOI 10.1007/978-3-319-16507-3

Springer Cham Heidelberg New York Dordrecht London
© Springer International Publishing Switzerland 2015
Softcover reprint of the hardcover 1st edition 2015

Printed on acid-free paper

Springer International Publishing AG Switzerland is part of Springer Science+Business Media
(www.springer.com)

Preface

This book evolved from a series of Ph.D. seminars presented to students both at Tsinghua University in Beijing and at the University of Illinois—Chicago. My lectures, in turn, grew out of my earlier investigation into sampling requirements in the sort of latent variable structural models that are prominent in social science fields like psychology, marketing, and information systems. In pursuit of my research, I discovered an academic domain steeped in mythologies and ambiguities, many that seem to have arisen during the fierce academic battles surrounding the Cowles Commission and Economics faculty at the University of Chicago in the 1940s. My home base in Chicago provided a unique opportunity to track down some of the original materials surrounding these debates. In the process, I gained an appreciation for the conflicting claims made by Cowles' systems of regression researchers, in contrast to those asserted by the isolated clique of Nordic researchers who single-handedly created the PLS path analysis and LISREL methodologies. The Cowles Commission ultimately prevailed in mainstream statistics and econometrics, but the Nordic approaches survived, and indeed thrived, in less rigorous arenas. This book surveys the full range of available structural equation modeling path analysis methodologies, from their early development in genetics to their current merging into network analysis tools. Applications of path analysis with structural equation models have steadily expanded over a broad range of disciplines—especially in the social sciences where many if not most key concepts are not directly observable, and the latent variable characteristics of these methods are especially desirable. This is the first book to extensively review the historical underpinnings that have defined the applications, methods, and assumptions invoked in each competing approach to estimation. History matters in understanding why particular academic disciplines have clustered around one or the other of these competing approaches. Knowing the background of PLS path analysis or LISREL is essential to understanding their strengths and weaknesses. The ability to accommodate unobservable theory constructs through latent variables has in particular grown with the expansion of the social sciences in universities. Latent variable constructs are fully explained here, and new methods are presented for extending their power. New techniques for path

analysis are surveyed along with guidelines for data preparation, sample size calculation, and the special treatment of Likert scale data. Tables of software, methodologies, and fit statistics provide a concise reference for any research program, helping
assure that its conclusions are defensible and publishable. It is my hope that both
scholars and students will find this book an accessible and essential companion to
their research.

Chicago, IL, USA   J. Christopher Westland
February 2015

Contents

Chapter 1
An Introduction to Structural Equation Models

The past two decades have witnessed a remarkable acceleration of interest in structural equation modeling (SEM) methods in many areas of research. In the social sciences, researchers often distinguish SEM approaches from more powerful systems of regression equation approaches by the inclusion of unobservable constructs (called latent variables in the SEM vernacular), and by the use of computationally intensive iterative searches for coefficients that fit the data. The expansion of statistical analysis to encompass unmeasurable constructs using SEM, canonical correlation, Likert scale quantification, principal components, and factor analysis has vastly extended the scope and relevance of the social sciences over the past century. Subjects that were previously the realm of abstract argumentation have been transported into the mainstream of scientific research.

The products of SEM statistical analysis algorithms fall into three groups: (1) pairwise canonical correlations between pairs of prespecified latent variables computed from observable data (from the so-called partial least squares path analysis, or PLS-PA approaches); (2) multivariate canonical correlation matrices for prespecified networks of latent variables computed from observable data (from a group of computer-intensive search algorithms originating with Karl Jöreskog); and (3) systems of regression approaches that fit data to networks of observable variables. A fourth approach is fast emerging with the introduction of powerful new social network analysis tools. These allow both visualization and network-specific statistics that draw on an old and rich literature in graph theory and physical network effects.

Most of the PLS-PA algorithms are variations on an incompletely documented software package released in 1980 (Lohmöller, 1988, 1989) and some even use this 40-year-old Fortran code unmodified inside a customized user interface wrapper. To make matters worse PLS-PA is a misnomer—something its inventor Herman Wold tried unsuccessfully to correct—and is unrelated to Wold's (1973) partial least squares regression methods, instead being a variation on Wold's (1966, 1975) canonical correlation methods.

Two different covariance structure algorithms are widely used: (1) LISREL (an acronym for LInear Structural RELations) (K. G. Jöreskog, 1970, 1993; Jöreskog &

© Springer International Publishing Switzerland 2015
J.C. Westland, *Structural Equation Models*, Studies in Systems,
Decision and Control 22, DOI 10.1007/978-3-319-16507-3_1

Sörbom, 1982; Jöreskog & Van Thillo, 1972; Jöreskog, Sorbom, & Magidson, 1979) and the AMOS (Analysis of Moment Structures) (Fox 2002, 2006; McArdle, 1988; McArdle & Epstein, 1987; McArdle & Hamagami, 1996, 2001). Variations on these algorithms have been implemented in EQS, TETRAD, and other packages.

Methods in systems of equation modeling and social network analytics are not as familiar in the social sciences as the first two methods, but offer comparatively more analytical power. Accessible and comprehensive tools for these additional approaches are covered in this book, as are research approaches to take advantage of the additional explanatory power that these approaches offer to social science research.

The breadth of application of SEM methods has been expanding, with SEM increasingly applied to exploratory, confirmatory, and predictive analysis through a variety of ad hoc topics and models. SEM is particularly useful in the social sciences where many if not most key concepts are not directly observable, and models that inherently estimate latent variables are desirable. Because many key concepts in the social sciences are inherently latent, questions of construct validity and methodological soundness take on a particular urgency. The popularity of SEM path analysis methods in the social sciences in one sense reflects a more holistic, and less blatantly causal, interpretation of many real-world phenomena than is found in the natural sciences. Direction in the directed network models of SEM arises from presumed cause-effect assumptions made about reality. Social interactions and artifacts are often epiphenomena—secondary phenomena that are difficult to directly link to causal factors. An example of a physiological epiphenomenon is, for example, time to complete a 100-m sprint. I may be able to improve my sprint speed from 12 to 11 s—but I will have difficulty attributing that improvement to any direct causal factors, like diet, attitude, and weather. The 1-s improvement in sprint time is an epiphenomenon—the holistic product of interaction of many individual factors. Such epiphenomena lie at the core of many sociological and psychological theories, and yet are impossible to measure directly. SEM provides one pathway to quantify concepts and theories that previously had only existed in the realm of ideological disputations.

To this day, methodologies for assessing suitable sample size requirements are a worrisome question in SEM-based studies. The number of degrees of freedom in structural model estimation increases with the number of potential combinations of latent variables, while the information supplied in estimating increases with the number of measured parameters (i.e., indicators) times the number of observations (i.e., the sample size)—both are nonlinear in model parameters. This should imply that requisite sample size is *not* a linear function solely of indicator count, even though such heuristics are widely invoked in justifying SEM sample size. Monte Carlo simulation in this field has lent support to the nonlinearity of sample size requirements. Sample size formulas for SEM are provided in the latter part of this book, along with assessments of existing rules of thumb. Contrary to much of the conventional wisdom, sample size for a particular model is constant across methods— PLS-PA, LISREL, and systems of regression approaches all require similar sample sizes when similar models are tested at similar power and significance levels. None of these methods generates information that is not in the sample, though at the margin particular methods may more efficiently use sample information in specific situations.

1.1   The Problem Domains of Structural Equation Models

Many real-world phenomena involve networks of theoretical constructs of interest to both the natural and social sciences. Structural equation modeling has evolved to help specify real-world network models to fit observations to theory. Early approaches lacked the computational power to do little more that trace out pathways along the networks under study. These so-called path models were initially applied in the natural sciences to map networks of heritable genetic traits: constructs such as black hair, long ears, and so forth in laboratory animals, with relationship links defined by ancestry. This early research was directed towards developing useful models of inheritance from straightforward observation, without the benefit of pre-existing theories. In case a model did not at first fit the data, researchers had easy access to additional observations that theoretically could be replicated without end, simply by breeding another generation.

Quantification of the social sciences during the mid-twentieth century demanded statistical methods that could assess the abstract and often unobservable constructs inherent in these softer research areas. Many social science observations, e.g., a year of economic performance in the US economy, could never be repeated or replicated. Data was the product of quasi-experiments: non-replicable, with potential biases controlled via expanded scope rather than replication of the experiment. Early work empirically tested pairwise relationships between soft constructs using canonical correlations (Hotelling, 1936). Structural equation methods for the social sciences grew out of the empirical quantification of social research constructs, with pioneering work by Rensis Likert in psychology, politics, and sociology (R. Likert, 1932; Rensis Likert, Roslow, & Murphy, 1934; Quasha & Likert, 1937), Edward Deming (Stephan, Deming, & Hansen, 1940) in national census statistics and Lee Cronbach in education (Cronbach & Meehl, 1955). Pursuit of network models in the social sciences fostered the development of SEM models paralleling economic's systems of regression approaches, but supporting the empirical assessment of networks of unobservable constructs. These SEM approaches (PLS-PA and LISREL) evolved in the social sciences from statistical methods in canonical correlation, and were designed to fit data to networks of unobservable constructs. Herman Wold (1966, 1973, 1974, 1975) and his student, former high school teacher Karl Jöreskog (K. G. Jöreskog & Van Thillo, 1972), were focused on hypothesis testing simple theories about the structural relationships between unobservable quantities in social sciences. Applications started with economics, but found greater usefulness in measuring unobservable model constructs such as intelligence, trust, value, and so forth in psychology, sociology, and consumer sentiment. These approaches remain popular today, as many central questions in the social sciences involve networks of abstract ideas that are often hard to measure directly.

Studies in economics and finance were less inclined towards abstractions, rather were challenged with the analysis and understanding of masses of dollar-denominated financial data. In this more tangible world, systems of regression models borrowed heavily from methods for analysis of transportation networks. These were employed to expand and extend existing linear regression modeling approaches (which themselves had been coopted from the astronomers). Systems of regression equation approaches pioneered by Tjalling Koopmans (1951, 1957, 1963) were designed to *prove or disprove theories about the structural relationships between economic measurements*. Because regression fits actual observations rather than abstract concepts, they are able to provide a wealth of goodness-of-fit information to assess the quality of the theoretical models tested. Wold and Jöreskog's methods provide goodness-of-fit information that in comparison is sparse, unreliable, and difficult to interpret.

Over the past two decades, many computer-intensive approaches have been developed to specify network models from data and to simulate observed behavior of real-world networks. Social network analysis extends graph theory into empirical studies. It takes observations (e.g., Wright's genetic traits) and classifies them as "nodes" (also called "vertices"). It infers relationships, called "edges" or "links" between these nodes from real-world observations. Social networks reflect social relationships in terms of individuals (nodes) and relationships (links) between the individuals. Examples of links are contracts, acquaintances, kinship, employment, and romantic relationships.

1.2   Motivation and Structure of This Book

The performance and behavioral characteristics of the three main structural equation model approaches have been reasonably well understood since their inception in the 1960–1970s. Unfortunately, for whatever reason, their strengths and weaknesses are too often misrepresented, or worse obscured behind sometimes specious fit statistics. The remainder of this book is designed to help the reader better understand what is happening on "the path"—specifically, how should a path coefficient be interpreted in a structural equation model. I try to keep things simple—my examples have purposely been limited to the simplest models possible, typically involving only two latent variables and one path. I also offer a comprehensive and accessible review of the current state of the art that will allow researchers to maximize research output from their datasets and confidently use available software for data analysis. The reader will find that the mathematics have purposely been kept to a minimum. The availability of extensive computer software, many with well-developed graphical interfaces, means that researchers can focus their efforts on assuring that research questions, design, data collection, and interpretation are accurate and complete while leaving the technical intricacies of data selection, tabulation, and statistical calculation to the software.

This book itself came together in small parts over a period of about 7 years. During those years—in my editorial roles at a number of journals—I reviewed many papers invoking path analysis methods, and repeatedly saw the same errors. Many of these problems arise because SEM software is now so well developed. Modern SEM software makes it possible to crank out results for complex, networked models that are based on only the vaguest of intuition, and then to test these same intuitions with data that does not directly measure them—and to do it all without understanding the statistics. In the wrong hands, this is surely a recipe for bad science. Though I will not dwell on research failure, I will occasionally cite cases where misuse of SEM has generated bad science, and what the consequences of these failures have been.

I hope that this book can provide a guide to generating reliable, defensible, and publishable answers to research questions, without the researcher expending an inordinate amount of time and effort trying to understand the philosophical debates and mathematical intricacies that allow us to confidently apply the techniques of SEM path analysis. In that sense, this book treats SEM as a "black box"—a box festooned with cranks, dials, knobs, and input readers. The researcher needs to know the consequences of particular controls and settings, and the meaning of particular dials, but need not understand the inner workings of the box. When driving a car, it is safer to spend all one's attention making corrections for road situations than to worry about engine timing; the analogy carries over to SEM path analysis.

Chapter 2 surveys the development of structural equation models—from its roots in genetics, and the concessions made to accommodate the labor-intensive manual computation of the day, to the computationally intensive approaches that proliferate today. I attempt to accurately portray the trade-offs and controversies that each approach engendered along the way.

Chapter 3 traces the statistical debates that evolved in the 1950s as the USA and the World were seeking out national economic and social statistics, and the development of early partial least square path analysis (PLS-PA) methods with canonical correlations. Prior to WWII there were few if any national statistics to guide policy making at a national level. The USA experienced an unprecedented period of ferment in conception and development of social and economic statistics in the late 1940–1950s (Karabell, 2014). Challenged to find new ways to substantiate the abstract ideas that drove spending on social security, education, banking, and numerous other government programs, statisticians derived latent variable statistical approaches, approaches that fit whole networks of metrics, and methods that took advantage of the rapidly developing computing industry.

Chapter 4 surveys one offshoot of canonical correlation network analysis called LInear Structural RELations or LISREL for short. There have been many variations on LISREL (AMOS, EQS, TETRAD) providing the same results with slightly different algorithms. Chapter 5 looks at another offshoot of network analysis that added linear regression (originally developed in astronomy) to analytics for transportation networks. These systems of regression equation approaches became standard analytical tools in the economics community.

Chapters 6 and 7 explain the role of data in model building, testing, and inference, and offer sample size formulas and explanations of procedures for differing types and distributions of data. Chapter 6 covers the standard parametric distributions, and Chap. 7 focuses on Likert scale survey data to which SEM is commonly applied.

Chapter 8 explores the many aims that would motivate a researcher to use SEM in addressing a research question. Since many questions in social sciences can only be addressed through individual perceptions, impressions, and judgments, questions of objective truth are generally somewhat elusive. Questions involving a consumer's willingness to pay for a product or service, for example, can be measured prior to purchase by the stated buying intentions of a group of consumers; or they could be measured after purchase by the money spent: both are noisy signals about "willingness." Correctly and reliably answering research questions give statisticians insight into the "true state of nature," into the real world, and into truth.

Chapter 9 surveys the emerging successors to structural equation modeling in the social sciences that are classified generally under the rubric of social network analysis. The paths of structural equation models were always a kludge, a hodgepodge of early twentieth-century statistical techniques cobbled together to help explore naturally occurring networks. Scientists in the past simply did not possess the analytical tools to map more than a few links at a time; path models were the best we could do at the time. But modern approaches are harnessing the power of advanced processors, cloud storage, and sophisticated algorithms to finally make possible research paradigms about which turn of the twentieth-century researchers could only dream.

The remainder of this book is designed to help the reader better understand what is happening on "the path"—specifically, how should a path coefficient be interpreted in a structural equation model. I try to keep things simple—my examples have purposely been limited to the simplest models possible, typically involving only two latent variables and one path. I also offer a comprehensive and accessible review of the current state of the art that will allow researchers to maximize research output from their datasets and confidently use available software for data analysis. The reader will find that the mathematics have purposely been kept to a minimum. The availability of extensive computer software, many with well-developed graphical interfaces, means that researchers focus their efforts on assuring that research questions, design, data collection, and interpretation are accurate and complete while leaving the technical intricacies of data selection, tabulation, and statistical calculation to the software.

I hope that this book can provide a guide to generating reliable, defensible, and publishable answers to research questions, without the researcher expending an inordinate amount of time and effort trying to understand the philosophical debates and mathematical intricacies that allow us to confidently apply the techniques of SEM path analysis. The researcher needs to know the consequences of particular controls and settings, and the meaning of particular dials, but need not understand the inner workings of the box. By the end of this book, aspiring researchers should have a very good grasp of the knobs, dials, and controls available in both existing and emerging methods in SEM and network analysis.

References

Cronbach, L. J., & Meehl, P. E. (1955). Construct validity in psychological tests. *Psychological Bulletin, 52*(4), 281.

Fox, J. (2006). Structural equation modeling with the SEM package in R. *Structural Equation Modeling, 13*(3), 465–486.

Fox, J. (2002). Structural equation models. *CRAN website*.

Hotelling, H. (1936). Relations between two sets of variates. *Biometrika, 28*(3/4), 321–377.

Jöreskog, K. G. (1970). A general method for estimating a linear structural equation system. *ETS Research Bulletin Series, 1970*, i–41.

Jöreskog, K. G. (1993). Testing structural equation models. In K. A. Bollen & J. S. Long (Eds.), *Sage focus editions* (Vol. 154, p. 294). Thousand Oaks, CA: Sage.

Jöreskog, K. G., & Sörbom, D. (1982). Recent developments in structural equation modeling. *Journal of Marketing Research, 19*, 404–416.

Jöreskog, K. G., Sorbom, D., & Magidson, J. (1979). *Advances in factor analysis and structural equation models*. Cambridge, MA: Abt books.

Jöreskog, K. G., & Van Thillo, M. (1972). LISREL: a general computer program for estimating a linear structural equation system involving multiple indicators of unmeasured variables. *ETS Research Bulletin Series, 1972*, i–72.

Karabell, Z. (2014). *The leading indicators: a short history of the numbers that rule our world*. New York, NY: Simon and Schuster.

Koopmans, T. C. (1951). Analysis of production as an efficient combination of activities. In *Activity analysis of production and allocation* (Vol. 13, pp. 33–37).

Koopmans, T. C. (1957). *Three essays on the state of economic science* (Vol. 21). New York, NY: McGraw-Hill.

Koopmans, T. C. (1963). *Appendix to 'On the Concept of Optimal Economic Growth': Cowles Foundation for Research in Economics*. New Haven, CT: Yale University.

Likert, R. (1932). A technique for the measurement of attitudes. *Archives of Psychology, 22*, 1–55.

Likert, R., Roslow, S., & Murphy, G. (1934). A simple and reliable method of scoring the Thurstone attitude scales. *The Journal of Social Psychology, 5*(2), 228–238.

Lohmöller, J.-B. (1988). The PLS program system: latent variables path analysis with partial least squares estimation. *Multivariate Behavioral Research, 23*(1), 125–127.

Lohmöller, J.-B. (1989). *Latent variable path modeling with partial least squares*. Heidelberg: Physica-Verlag.

McArdle, J. J. (1988). Dynamic but structural equation modeling of repeated measures data. In J. R. Nesselroade & R. B. Cattell (Eds.), *Handbook of multivariate experimental psychology* (pp. 561–614). New York, NY: Springer.

McArdle, J. J., & Epstein, D. (1987). Latent growth curves within developmental structural equation models. *Child Development, 58*, 110–133.

McArdle, J. J., & Hamagami, F. (1996). Multilevel models from a multiple group structural equation perspective. In G. A. Marcoulides & R. E. Schumacker (Eds.), *Advanced structural equation modeling: issues and techniques* (pp. 89–124). Mahwah, NJ: Erlbaum.

McArdle, J. J., & Hamagami, F. (2001). Latent difference score structural models for linear dynamic analyses with incomplete longitudinal data. In L. M. Collins & A. G. Sayer (Eds.), *New methods for the analysis of change decade of behavior* (pp. 139–175). Washington, DC: US American Psychological Association.

Quasha, W. H., & Likert, R. (1937). The revised Minnesota paper form board test. *Journal of Education and Psychology, 28*(3), 197.

Stephan, F. F., Deming, W. E., & Hansen, M. H. (1940). The sampling procedure of the 1940 population census. *Journal of the American Statistical Association, 35*(212), 615–630.

Wold, H. (1966). Estimation of principal components and related models by iterative least squares. *Multivariate Analysis, 1*, 391–420.

Wold, H. (1973). Nonlinear iterative partial least squares (NIPALS) modelling: some current developments. *Multivariate Analysis, 3*, 383–407.

Wold, H. (1974). Causal flows with latent variables: partings of the ways in the light of NIPALS modelling. *European Economic Review, 5*(1), 67–86.

Wold, H. (1975). *Path models with latent variables: the NIPALS approach.* New York, NY: Academic Press.

Chapter 2
A Brief History of Structural Equation Models

Though structural equation models today are usually associated with soft problems in the social sciences, they had their origin in the natural sciences—specifically biology. Europe's nineteenth-century scholars were challenged to make sense of the diverse morphologies observed during an age of explorations, in Asia, Africa, and the Americas, as well as at home. In this period, new species of plants and animals were transplanted, domesticated, eaten, and bred at an unprecedented rate. An American ultimately provided one statistical tool that allowed scholars to build a science out of their diverse observations.

2.1 Path Analysis in Genetics

Seldom has a nonhuman animal been so thoroughly poked, observed, trained, and dissected as the domesticated dog. A member of the *Canidae* family, the dog is distantly related to coyotes and jackals, dingoes and dholes, foxes and wolves. There is evidence of distinct dog breeds as early as 5,000 years ago in drawings from ancient Egypt. The business of designing dogs for particular purposes began in earnest around the sixteenth century, and by the nineteenth century, clubs and competitions abounded for the naming and monitoring of breeds. There is a huge variation of sizes, shapes, temperaments, and abilities in modern dogs—much more so that in their homogeneous wolf ancestors. This has resulted from humans consciously influencing the genetics of dog populations through an involved network of interbreeding and active selection.

But none of this was a science at the dawn of the twentieth century, despite enormous expenditures, and centuries of breeding and contests to create "the perfect dog." There was no theory (or perhaps too many competing but unsupported theories) about how particular characteristics arose in a particular subpopulation of dogs.

© Springer International Publishing Switzerland 2015
J.C. Westland, *Structural Equation Models*, Studies in Systems,
Decision and Control 22, DOI 10.1007/978-3-319-16507-3_2

The sciences of evolution and genetics seldom spoke to each other before the twentieth century. The most influential biologists held the idea of blending inheritance, promoted in a particular form in Charles Darwin's theory of pangenesis—inheritance of tiny heredity particles called gemmules that could be transmitted from parent to offspring. In those days, the work of the Augustinian friar and polymath Gregor Mendel was unknown, having been rejected and forgotten in the biology community when published in the 1860s. Mendel's sin was to introduce mathematics into a field that biologists felt should be a descriptive science, not an analytical one.

Rediscovery of Mendel's writings in the early twentieth century led biologists towards the establishment of genetics as a science and basis for evolution and breeding. Geneticist, Sewall Wright, along with statisticians R. A. Fisher and J. B. S. Haldane, was responsible for the modern synthesis that brought genetics and evolution together.

Wright's work brought quantitative genetics into animal and plant breeding, initiating the hybrid seed revolution that transformed US agriculture in the first half of the twentieth century. Wright actively mapped the breeding networks that created desirable hybrids. Of particular significance to the dog breeders was Wright's discovery of the inbreeding coefficient and of methods of computing it in pedigrees.

The synthesis of statistical genetics into the evolution of populations required a new quantitative science with which to map the networks of influence, on random genetic drift, mutation, migration, selection, and so forth. Wright's quantitative study of influence networks evolved in the period 1918 through 1921 into Wright's statistical method of path analysis—one of the first statistical methods using a graphical model, and one which is the subject of this book.

Let's begin by reviewing the evolution of path analysis from the dark ages of nineteenth-century evolution debates, through today's statistical methods, to emerging techniques for mapping the extensive networks of biological interactions important to genetics and biotechnology in the future.

2.2   Sewall Wright's Path Analysis

Path analysis was developed in 1918 by geneticist Sewall Wright (1920, 1921, 1934) who used it to analyze the genetic makeup of offspring of laboratory animals (Fig. 2.1).

Early graphs were very descriptive, with pictures and stories attached. But gradually pictures of laboratory critters gave way to representative boxes and positive or negative correlations (Fig. 2.2).

Rensis Likert's work at the University of Michigan in the 1930s and 1940s saw path analysis directed towards social science research. Social scientists need to model many abstract and unobservable constructs—things like future intentions, happiness, customer satisfaction, and so forth. Though not directly observable, there typically exist numerous surrogates that can provide insight into such abstract

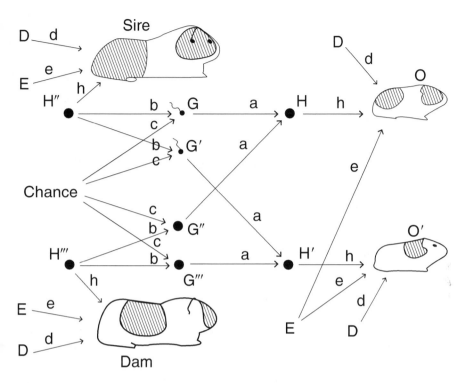

Fig. 2.1 Relations between littermates (0,0) and between each of them and their parents. *H*, *H'*, *H''*, and *H'''* represent the genetic constitutions of the four individuals; *G*, *G'*, *G''*, and *G'''* that of four germ cells. *E* represents such environmental factors as are common to littermates. *C* represents other factors, largely ontogenetic irregularity. The *small letters* stand for the various path coefficients (Wright, 1920)

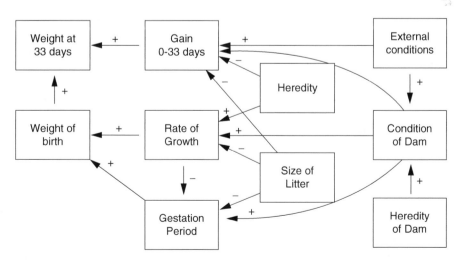

Fig. 2.2 Interrelations among the factors which determine the weight of guinea pigs at birth and at weaning (Wright, 1921)

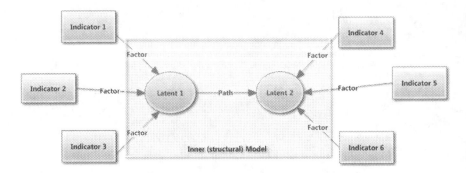

Fig. 2.3 A generic path model with latent variables

(or latent) constructs—these observable surrogates are called "indicators" of the latent variable.

Further innovation in path models evolved around Hermann Wold's extensions of Hotelling's seminal work in principal component analysis (PCA). Wold began promoting the principal components as representations of abstract (latent) constructs. Latent abstractions proved useful in the evolving fields of psychometrics and sociological surveys, and were widely adopted in the 1950s and 1960s (Hotelling, 1936; Wold, 1966).

Path diagrams evolved once again, to incorporate Wold's conceptualization of latent constructs as the first component from a PCA. Wold called the network model of latent variables the "structural model" or sometimes the "inner" model. The term "structural equation model" came about from his use, which Wold borrowed from the matrix terminology of systems of equation regression approaches developed at the Cowles Commission.

Social scientists were ultimately not content to let PCA dictate their choice of abstractions. In education research, Henry Kaiser and Lee Cronbach, both faculty in the University of Illinois, School of Education, in the 1950s argued that such abstract concepts could be conceived prior to data collection, and the collected data with the abstract concept could be reviewed after the fact to see that it actually looks like a first principal component.

These alternative approaches to defining the relationship between indicators and the latent variables they indicate created what Wold called formative and reflective links. If the researcher chooses the indicators before the latent variables, the links are called "formative" because the factor (link) weights and the latent variable are formed from the first component of the PCA. If the researcher chooses the latent construct before the indicators, the links are called "reflective" because the factor (link) weights are believed to reflect the abstract (latent) construct.

By the 1960s ad hoc path diagrams had evolved to neat boxes and bubbles that identified measurable and latent components of the model (Fig. 2.3).

2.3   Networks and Cycles

Many real-world influences are basically cyclic, and interesting questions revolve around the convergence to equilibrium—e.g., in predator–prey ratios and corporate "satisficing".

Path tracing has become an essential feature of the graphical interface for SEM software programs. These specific rules are designed to yield graphs (and thus models) that are non-recursive—i.e., do not have influence loops. Consider, for example, a graph of three variables A, B, and C with a recursive relationship (Fig. 2.4):

Assume that the correlation between each pair of latent variables in the figure is 0.5; thus a change in one variable results in a linear influence of 50 % of that change on the variable that the outgoing arrow points to. Then we might ask "what will be the net effect on all variables (including A) if we vary A by 1.0?" The variance of A will be affected by the initial perturbation; 50% of that will appear at B, 25 % at C, 12.5 % at A again, and so forth. This is not a result that can be teased out with regressions (nor with PLS path analysis).

The expected correlation due to each chain traced between two variables is the product of the standardized path coefficients, and the total expected correlation between two variables is the sum of these contributing path chains. Intrinsically, Wright's rules assume a model without feedback loops. It puts paid to the mental simplification of a simple linear sequence of causal pathways.

Path modeling software will generally not allow the design of network graphs with cycles—the graphs will have to be acyclic. In order to validly calculate the relationship between any two boxes Wright proposed a simple set of path tracing

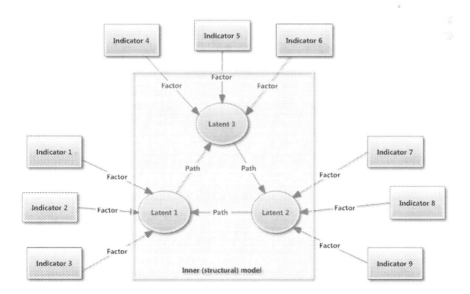

Fig. 2.4 A cyclic graph (recursive)

rules for calculating the correlation between two variables (Wright, 1934). The correlation is equal to the sum of the contribution of all the pathways through which the two variables are connected. The strength of each of these contributing pathways is calculated as the product of the path coefficents along that pathway.

The rules for path tracing are based on the principle of never allowing flow to pass out of one arrowhead and into another arrowhead. These are the following:

1. You can trace backward up an arrow and then forward along the next, or forward from one variable to the other, but never forward and then back.
2. You can pass through each variable only once in a given chain of paths.
3. No more than one bidirectional arrow can be included in each path chain.

These three rules assured Wright that he would estimate paths on a directed acyclic graph, that is, a network model formed by a collection of vertices and directed edges, each edge connecting one vertex to another, such that there is no way to start at some vertex and follow a sequence of edges that eventually loops back to that vertex again. Directed acyclic graphs pose significantly fewer problems to mathematical analysis, and restricting path analysis allows simpler calculations, with little loss of generality. Introduction of computer-intensive analysis of variance approaches by Jöreskog (1967, 1970) in the 1970s allowed more general network models ultimately to be estimated.

2.4   What Is a Path Coefficient?

Wright was satisfied estimating his paths with correlation coefficients. This made a great deal of sense before the era of modern computers (or even hand calculators). Furthermore, he argued that in many cases, any other path coefficients required could be recovered from transformations of these correlation coefficients (Wright 1960). Nonetheless, as path modeling grew more common, statisticians began experimentation with covariance and regression coefficients. This evolution started with a suggestion in the early 1950s. Tukey (1954) advocated systematic replacement in path analysis of the dimensionless path coefficients by the corresponding concrete path regressions. Geneticists Turner and Stevens (1959) published the seminal paper presenting the mathematics of path regression and inviting extensions that would apply other developments in statistics. In the early days of data processing, both Hermann Wold and his student Karl Jöreskog developed software adopting variations of Turner and Stevens' mathematics that took advantage of computer-intensive techniques developed in the 1970s and 1980s. Over time, the candidate list for path coefficients has steadily broadened. Here is a short list of some of the alternatives that are popular:

1. Pearson correlation (Wright, 1921)
2. Canonical correlation between latent variables (Jöreskog, 1967; Jöreskog & Van Thillo, 1972; Wold, 1966, 1974)
3. Regression coefficients (Wold, 1975)
4. Covariance (Jöreskog & Van Thillo, 1972)

5. Systems of equation regression coefficient (C. A. Anderson, 1983; Zellner, 1962; Zellner & Theil, 1962)
6. Generalized distance measures (A. L. Barabási, Dezső, Ravasz, Yook, & Oltvai, 2003; A. L. Barabási & Oltvai, 2004)

The work of Tukey and Turner and Stevens along with the burgeoning availability of scientific computers in universities ushered in an era of ferment, where computationally intensive methods were invoked to develop evermore complex (and possibly informative) path coefficients.

2.5 Applications and Evolution

Geneticist Sewall Wright was extremely influential, and in great demand as a reviewer and editor. His work was widely cited and served as a basis for a generation of statistical analysis in genetic and population studies in the natural sciences. Wright's work influenced early 1940s' and 1950s' Cowles Commission work on simultaneous equation estimation centered on Koopmans' algorithms from the economics of transportation and optimal routing. This period witnessed the development of many of the statistical techniques we depend on today to estimate complex networks of relationships between both measured (called factors or exogenous, manifest, or indicator variables) and imagined variables (called latent or endogenous variables). After Sewall Wright's seminal work, structural equation models evolved in three different streams: (1) systems of equation regression methods developed mainly at the Cowles Commission; (2) iterative maximum likelihood algorithms for path analysis developed mainly at the University of Uppsala; and (3) iterative least squares fit algorithms for path analysis also developed at the University of Uppsala. Figure 2.5 describes the pivotal developments in latent variable statistics in terms of method (precomputer, computer intensive, and a priori SEM) and objectives (exploratory/prediction or confirmation).

Both LISREL (an acronym for *linear structural relations*) and partial least squares (PLS) were conceived as computer implementations, with an emphasis from the start on creating an accessible graphical and data entry interface—and extension of Sewell Wright's path analysis diagrams—to ensure widespread usage. Additionally they were designed to incorporate "latent factors" following the example of three other multivariate alternatives which involved latent constructs: (1) *discriminant analysis*—the prediction of group membership from the levels of continuous predictor variables; (2) *principal component regression*—the prediction of responses on the dependent variables from factors underlying the levels of the predictor variables; and (3) *canonical correlation*—the prediction of factors underlying responses on the dependent variables from factors underlying the levels of the predictor variables.

In an interesting historical footnote, the application of computers to Wold's partial least squares regression (PLSR) version of path analysis was short lived. PLSR computes regression coefficients where data are highly multicollinear (and finds modern use in spectrographic analysis). But this is seldom a problem in path

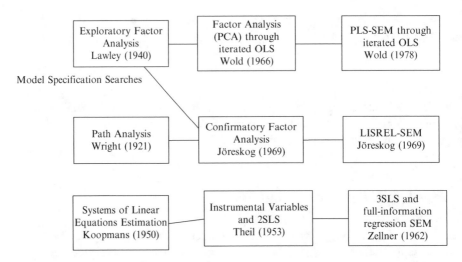

Fig. 2.5 History of SEM statistical methods and their precursors

analysis, and when Wold's research associate Lohmöller released his desktop computing software LVPLS that implemented both Jöreskog's as well as his own interpretation of Wold's ideas for path modeling, he implemented ordinary least squares for computing path coefficients, because it was faster to compute, and yielded the same coefficient values. Lohmoller's algorithms (Lohmöller, 1981) are used in all of the modern PLS path analysis packages. So even though PLS path analysis includes PLS in the name, it does not use PLSR in its algorithm. Wold actually tried to change the acronym PLS, but the name stuck.

2.6 The Chicago School

Regression methods date back to the early nineteenth century, when Legendre (1805) and Gauss (1809) developed least squares regression, correlation methods, eigenvalues, eigenvectors, determinants, and matrix decomposition methods. All were extensions of Gauss' theory of algebraic invariants, used initially to fit data on orbits of astronomical objects, where much was known about the nature of errors in the measurements and in the equations, and where there was ample opportunity for comparing predictions to reality. Astronomer and social scientist Adolphe Quetelet (1835) was among the first who attempted to apply the new tools to social science, planning what he called a "social physics," a discipline that was well evolved by the late nineteenth century. The resulting tools provided the basis for what would eventually become the study of econometrics.

Alfred Cowles made his fortune in the insurance industry in the 1920s, and managed to keep that fortune intact after 1929. During a lengthy hospital stay, he started collecting financial data (difficult in the days before the SEC and auditing), a project which ultimately grew into the modern Compustat/CRSP databases. In

1932 he used his fortune to set up the Cowles Commission (first at Cowles' alma mater Yale, later moving to Chicago when Nobelist Tjalling Koopmans moved, and later back to Yale) to develop econometric statistical models for his databases. From his own experience in investment counseling, he had been frustrated by the "guessing game" techniques used in forecasting the stock market and believed that scholarly research by experts could produce a better method. He suggested as much in a famous book "Can Stock Market Forecasters Forecast?" (Cowles, 1933).

Early Cowles' work on simultaneous equation estimation centered on Tjalling Koopmans' algorithms from the economics of transportation and optimal routing. SEM work centered on maximum likelihood estimation, and closed form algebraic calculations, as iterative solution search techniques were limited in the days before computers. T. W. Anderson and Rubin (1949, 1950) developed the limited information maximum likelihood (LIML) estimator for the parameters of a single structural equation, which indirectly included the 2SLS estimator and its asymptotic distribution (T. W. Anderson 2005; T. W. Anderson, Kunitomo, & Matsushita, 2010; Farebrother, 1999). Two-stage least squares (2SLS) was originally proposed as a method of estimating the parameters of a single structural equation in a system of linear simultaneous equations, being introduced by Theil (1953a, 1961, 1992) and more or less independently by Basmann (1988) and Sargan (1958). Anderson's LIML was eventually implemented in a computer search algorithm, where it competed with other iterative SEM algorithms.

Two-stage least squares regression (2SLS) was by far the most widely used systems of equations method in the 1960s and the early 1970s. The explanation involves both the state of statistical knowledge among applied econometricians and the primitive computer technology available at the time. The classic treatment of maximum likelihood methods of estimation is presented in two Cowles Commission monographs: (Turner & Stevens, 1959) *Statistical Inference in Dynamic Economic Models*, and (Hood, Koopmans, & Economics, 1953) *Studies in Econometric Method*. By the end of the 1950s computer programs for ordinary least squares were available. These programs were simpler to use and much less costly to run than the programs for calculating LIML estimates or other approaches requiring iterative search for solutions. Owing to advances in computer technology, and, perhaps, also the statistical background of applied econometricians, the popularity of 2SLS started to wane towards the end of the 1970s. Computing advances meant that the difficulty of calculating LIML estimates was no longer a daunting constraint.

Zellner (Zellner, 1962; Zellner & Theil, 1962) algebraically extended 2SLS to a full information method using his *seemingly unrelated regressions technique*, calling the result three-stage least squares (3SLS). In general, the Chicago approaches were couched as systems of equations, without the familiar "arrow and bubble" diagrams we have come to associate with SEM. Geneticist Sewall Wright conducted a seminar in the 1940s on path coefficients to the Cowles Commission emphasizing the graphical "bubble and arrow" diagrams that he used in path analysis, and which have since become synonymous with path analysis and its extensions to SEM. Unfortunately, neither Wright nor the Cowles econometricians saw much merit in the other's methods (Christ, 1994) and the main body of Cowles research continued to be dominated by the algebra of Tjalling Koopmans' systems of equations and optimal routing perspective.

2.7   The Scandinavian School

Structural equation modeling (SEM) statistical methods provide statistical fitting to data of causal models consisting of unobserved variables. SEM approaches were computer-intensive attempts to generalize Sewall Wright's path analysis methods, motivated by the merging of path analysis with the systems of equation econometrics of the Cowles Commission. Sewall Wright was also involved in the development of factor analysis. Charles Spearman (Spearman, 1950) pioneered the use of factor analysis, discovering that school children's scores in various subjects were strongly correlated. He suggested that individual subject scores were indicators of a single factor representing general mental ability. Path analysis was applied elsewhere in sociology by Cowles and Chapman (1935) and psychology by Werts and Linn (Werts & Linn, 1970; Werts, Linn, & Jöreskog, 1974). The latter two studies introduced many of the new ideas from econometrics to areas where "latent" factors played a major part in theory. Ideas congealed gradually between the mid-1960s and the mid-1980s when most of the vocabulary we would recognize today was in place.

SEM path modeling approaches were developed at the Cowles Commission building on the ideas of the geneticist Wright, and championed at Cowles by Nobelist Trygve Haavelmo. Unfortunately, SEM's underlying assumptions were challenged by economists such as Freedman (1987) who objected that SEM's "failure to distinguish among causal assumptions, statistical implications, and policy claims has been one of the main reasons for the suspicion and confusion surrounding quantitative methods in the social sciences." Haavelmo's path analysis never gained a large following among US econometricians, but was successful in influencing a generation of Haavelmo's fellow Scandinavian statisticians. Hermann Wold (University of Uppsala) developed his Fixed-Point and PLS approaches to path modeling, and Karl Jöreskog (University of Uppsala) developed LISREL maximum likelihood approaches. Both methods were widely promoted in the USA by University of Michigan marketing professor Claes Fornell (Ph.D., University of Lund) and his firm CFI, which has conducted numerous studies of consumer behavior using SEM statistical approaches. Fornell introduced SEM techniques to many of his Michigan colleagues through influential books with David Larker (Fornell & Larcker, 1981) in accounting, Wynne Chin and Don Barclay in information systems (Barclay, Higgins, & Thompson, 1995; Chin, 1998; Chin & Newsted, 1999), Richard Bagozzi in marketing, and Fred Davis in validating the technology acceptance model in Davis, Bagozzi, and Warshaw (1989).

Development of SEM statistics in Sweden occurred in the late 1960s when computing power was just becoming widely available to academics, making possible computer-intensive approaches to parameter estimation. Jöreskog (1967, 1969, 1970) developed a rapidly converging iterative method for *exploratory* ML factor analysis (i.e., the factors are not defined in advance, but are discovered by exploring the solution space for the factors that explain the most variance) based on the Davidon-Fletcher-Powell math programming procedure commonly used in the solution of unconstrained nonlinear programs. As computing power evolved, other algorithms became feasible for searching the SEM solution space, and current software tends to use Gauss-Newton methods to optimize Browne's (Browne & Cudeck, 1989, 1992, 1993) discrepancy function with an appropriate weight matrix that converges to ML, ULS, or GLS solutions for the SEM or to Browne's asymptotically distribution-free discrepancy function using tetrachoric correlations.

Jöreskog (1969) extended this method to allow a priori specification of factors and factor loadings (i.e., the covariance of an unobserved factor and some observed "indicator") calling this *confirmatory* factor analysis. Overall fit of the a priori theorized model to the observed data could be measured by likelihood ratio techniques, providing a powerful tool for theory confirmation.

In work that paralleled Jöreskog's, Herman Wold (1966) described a procedure to compute principal component eigenvalues by an iterative sequence of OLS regressions, where loadings were identical to closed-form algebraic methods. In his approach the eigenvalues can be interpreted as the proportion of variance accounted for by the correlation between the respective "canonical variates" (i.e., the factor loadings) for weighted sum scores of the two sets of variables. These canonical correlations measured the simultaneous relationship between the two sets of variables, where as many eigenvalues are computed as there are canonical roots (i.e., as many as the minimum number of variables in either of the two sets). Wold showed that his iterative approach produced the same estimates as the closed-form algebraic method of Hotelling (1936). Lohmöller (1981) developed   PLS-PA computer software which generated a sequence of canonical correlations along paths on the network.

The PLS-PA designation has caused no end of confusion in the application of Lohmöller's software, which was casually and somewhat gratuitously called "partial least squares" as a marketing ploy. Wold was well known for his development of the entirely distinct partial least squares *regression* (PLSR) NIPALS algorithm. NIPALS provided an alternative to OLS using a design matrix of dependent and independent variables, rather than just the independent variables of OLS. PLSR tends to work well for multicollinear data, but otherwise offers no advantage over OLS.

Hauser and Goldberger (1971) were able to estimate a model of indicator (observed) variables and latent (unobserved) factors, with correlated indicators and error terms, using GLS; this work provided the basis for Wold's (1973, 1974, 1975) NIPALS algorithm through alternating iterations of simple and multiple OLS regressions. After Herman Wold's death in 1992, PLSR, which bears no relationship to PLS path analysis, continued to be promoted by his son, the chemist Svante Wold, through his firm *Umetrics*. Consequently, the most frequent application of PLSR is found in chemometrics and other natural sciences that generate large quantities of multicollinear data.

2.8   Limited and Full Information Methods

The search for estimators for simultaneous equation models can take place in one of the two ways: (1) "limited information" or path methods and (2) "full information" or network methods. Limited information methods estimate individual node pairs or paths in a network separately using only the information about the restrictions on the coefficients of that particular equation (ignoring restrictions on the coefficients of other equations). The other equations' coefficients may be used to check for identifiabilty, but are not used for estimation purposes. Full information methods estimate the full network model equations jointly using the restrictions on the parameters of all the equations as well as the variances and covariances of the residuals. These terms are applied both to observable and to latent variable models.

The most commonly used limited information methods are ordinary least squares (OLS), indirect least squares (ILS), two-stage least squares (2SLS), limited information maximum likelihood (LIML), and instrumental variable (IV) approaches. The OLS method does not give consistent estimates in the case of correlated residuals and regressors (which is commonly the situation in SEM analyses), whereas the other methods do provide consistent estimators. Yet OLS tends to be more robust with respect to model specification errors than the other limited information approaches, a problem that is exacerbated by small sample sizes. ILS gives multiple solutions, and thus is less favored than other approaches. The 2SLS approach provides one particular set of weightings for the ILS solutions; it also is a particular instrumental variable method. If the equations under consideration are exactly identified, then the ILS, 2SLS, IV, and LIML estimates are identical.

Partial least squares path analysis (PLS-PA) is also a limited information method. Dhrymes (Dhrymes, 1971a, 1971b; Dhrymes, Berner, & Cummins, 1974) provided evidence that (similar to Anderson's LIML) PLS-PA estimates asymptotically approached those of 2SLS with exactly identified equations. This in one sense tells us that with well-structured models, all of the limited information methods (OLS excluded) will yield similar results. We will revisit Dhryme's results when we discuss the behavior of PLS-PA estimators with varying sample sizes in the next chapter.

2.9   A Look Towards the Future

Path models were always a workaround—clumsy and inelegant, but used for lack of substitutes. Making sense of natural or man-made networks requires data and analytical power that, until recently, has not been available to researchers. PLS-PA, LISREL, and systems of regressions were designed for calculation on paper and with adding machines, barely wieldy given the size and complexity of the networks under analysis. Their conclusions have too often proven unreliable, simplistic, and inapplicable to prediction and understanding of the real world. Modern developments in social network analysis, borrowing heavily from graph theory, have moved forward our understanding of networks in the real world. Social network analysis is becoming key method in modern sociology, has found commercial use in Internet marketing, and is an emergent method in anthropology, biology, economics, marketing, geography, history, information systems, strategy, political science, psychology, and linguistics.

References

Anderson, C. A. (1983). The causal structure of situations: the generation of plausible causal attributions as a function of type of event situation. *Journal of Experimental Social Psychology, 19*(2), 185–203.
Anderson, T. W. (2005). Origins of the limited information maximum likelihood and two-stage least squares estimators. *Journal of Econometrics, 127*(1), 1–16.

Anderson, T. W., Kunitomo, N., & Matsushita, Y. (2010). On the asymptotic optimality of the LIML estimator with possibly many instruments. *Journal of Econometrics, 157*(2), 191–204.

Anderson, T. W., & Rubin, H. (1949). Estimation of the parameters of a single equation in a complete system of stochastic equations. *Annals of Mathematical Statistics, 20*(1), 46–63.

Anderson, T. W., & Rubin, H. (1950). The asymptotic properties of estimates of the parameters of a single equation in a complete system of stochastic equations. *Annals of Mathematical Statistics, 21*, 570–582.

Barabási, A.-L., Dezső, Z., Ravasz, E., Yook, S.-H., & Oltvai, Z. (2003). *Scale-free and hierarchical structures in complex networks.* Paper presented at the Modeling of Complex Systems: Seventh Granada Lectures.

Barabási, A. L., & Oltvai, Z. N. (2004). Network biology: understanding the cell's functional organization. *Nature Reviews Genetics, 5*(2), 101–113.

Barclay, D., Higgins, C., & Thompson, R. (1995). The partial least squares (PLS) approach to causal modeling: personal computer adoption and use as an illustration. *Technology Studies, 2*(2), 285–309.

Basmann, R.L. (1988). Causality tests and observationally equivalent representations of econometric models. *Journal of Econometrics 39*(1), 69–104.

Browne, M. W., & Cudeck, R. (1989). Single sample cross-validation indices for covariance structures. *Multivariate Behavioral Research, 24*(4), 445–455.

Browne, M. W., & Cudeck, R. (1992). Alternative ways of assessing model fit. *Sociological Methods & Research, 21*(2), 230–258.

Browne, M. W., & Cudeck, R. (1993). *Alternative ways of assessing model fit* (Sage focus editions, Vol. 154, p. 136). Thousand Oaks, CA: Sage.

Chin, W. W. (1998). Commentary: issues and opinion on structural equation modeling. *MIS Quarterly, 22*, vii.

Chin, W. W., & Newsted, P. R. (1999). Structural equation modeling analysis with small samples using partial least squares. In *Statistical strategies for small sample research* (Vol. 2, pp. 307–342). Thousand Oaks, CA: Sage.

Christ, C. F. (1994). The Cowles Commission's contributions to econometrics at Chicago, 1939–1955. *Journal of Economic Literature, 32*, 30–59.

Cowles, A. (1933). Can stock market forecasters forecast? *Econometrica, 1*, 309–324.

Cowles, A., 3rd, & Chapman, E. N. (1935). A statistical study of climate in relation to pulmonary tuberculosis. *Journal of the American Statistical Association, 30*(191), 517–536.

Davis, F. D., Bagozzi, R. P., & Warshaw, P. R. (1989). User acceptance of computer technology: a comparison of two theoretical models. *Management Science, 35*, 982–1003.

Dhrymes, P. J. (1971a). *Distributed lags.* San Francisco, CA: Holden-Day Inc.

Dhrymes, P. J. (1971b). Equivalence of iterative Aitken and maximum likelihood estimators for a system of regression equations. *Australian Economic Papers, 10*(16), 20–24.

Dhrymes, P. J., Berner, R., & Cummins, D. (1974). A comparison of some limited information estimators for dynamic simultaneous equations models with autocorrelated errors. *Econometrica, 42*, 311–332.

Farebrother, R. W. (1999). *Fitting linear relationships: a history of the calculus of observations 1750–1900.* New York, NY: Springer.

Fornell, C., & Larcker, D. F. (1981). Evaluating structural equation models with unobservable variables and measurement error. *Journal of Marketing Research, 18*, 39–50.

Freedman, D. A. (1987). As others see us: a case study in path analysis. *Journal of Educational and Behavioral Statistics, 12*(2), 101–128.

Gauss, K. F. (1809). Teoria motus corporum coelestium in sectionibus conicus solem ambientieum.

Goldberger, A. S., & Hauser, R. (1971). The treatment of unobservable variables in path analysis. *Sociological Methodology, 3*(8), 1.

Hood, W. C., Koopmans, T. C., & Cowles Commission for Research in Economics. (1953). *Studies in econometric method* (Vol. 14). New York, NY: Wiley.

Hotelling, H. (1936). Relations between two sets of variates. *Biometrika, 28*(3/4), 321–377.

Jöreskog, K. G. (1967). Some contributions to maximum likelihood factor analysis. *Psychometrika, 32*(4), 443–482.

Jöreskog, K. G. (1969). A general approach to confirmatory maximum likelihood factor analysis. *Psychometrika, 34*(2), 183–202.

Jöreskog, K. G. (1970). A general method for analysis of covariance structures. *Biometrika, 57*(2), 239–251.

Jöreskog, K. G., & Van Thillo, M. (1972). LISREL: a general computer program for estimating a linear structural equation system involving multiple indicators of unmeasured variables. *ETS Research Bulletin Series, 1972*, i–72.

Legendre, A. M. (1977). *Note par M.*** Second supplement to the third edition of Legendre (1805).* A separate pagination. English translation by Stigler. pp. 79–80.

Lohmöller, J. B. (1981). *Pfadmodelle mit latenten variablen: LVPLSC ist eine leistungsfähige alternative zu LIDREL.* München: Hochsch. d. Bundeswehr, Fachbereich Pädagogik.

Milgram, S. (1967). The small world problem. *Psychology Today, 2*(1), 60–67.

Quetelet, A. (1835). Sur l'homme et le développement de ses facultés ou essai de physique sociale. Bachelier, Paris.

Sargan, J. D. (1958). The estimation of economic relationships using instrumental variables. *Econometrica*, pp. 393–415.

Spearman, Charles, and Ll Wynn Jones. "Human ability." (1950).

Stigler, S. M. (1981). Gauss and the invention of least squares. *The Annals of Statistics*, pp. 465–474.

Theil, H. (1953). Repeated least squares applied to complete equation systems. Central Planning Bureau, The Hague.

Theil, H. (1992). Estimation and simultaneous correlation in complete equation systems. Henri Theil's contributions to economics and econometrics, pp. 65–107. Springer, Netherlands.

Theil, H. (1961). Economic forecasts and policy.

Tukey, J. W. (1954). Causation, regression, and path analysis. *Statistics and Mathematics in Biology*: 35–66.

Turner, M. E., & Stevens, C. D. (1959). The regression analysis of causal paths. *Biometrics, 15*(2), 236–258.

Werts, C. E., & Linn, R. L. (1970). Path analysis: psychological examples. *Psychological Bulletin, 74*(3), 193.

Werts, C. E., Linn, R. L., & Jöreskog, K. G. (1974). Intraclass reliability estimates: testing structural assumptions. *Educational and Psychological Measurement, 34*(1), 25–33.

Westland, J. C. (2010). Lower bounds on sample size in structural equation modeling. *Electronic Commerce Research and Applications, 9*(6), 476–487.

Wold, H. (1966). Estimation of principal components and related models by iterative least squares. *Multivariate Analysis, 1*, 391–420.

Wold, H. (1973). Nonlinear iterative partial least squares (NIPALS) modelling: some current developments. *Multivariate Analysis, 3*, 383–407.

Wold, H. (1974). Causal flows with latent variables: partings of the ways in the light of NIPALS modelling. *European Economic Review, 5*(1), 67–86.

Wold, H. (1975). *Path models with latent variables: the NIPALS approach.* New York, NY: Academic Press.

Wright, S. (1960). Path coefficients and path regressions: alternative or complementary concepts? Biometrics **16**(2), 189–202.

Wright, S. (1920). The relative importance of heredity and environment in determining the piebald pattern of guinea-pigs. *Proceedings of the National Academy of Sciences of the United States of America, 6*(6), 320.

Wright, S. (1921). Correlation and causation. *Journal of Agricultural Research, 20*(7), 557–585.

Wright, S. (1934). The method of path coefficients. *Annals of Mathematical Statistics, 5*(3), 161–215.

Wright, S. (1960). Path coefficients and path regressions: alternative or complentary concepts?. *Biometrics 16*(2), 189–202.

Zellner, A. (1962). An efficient method of estimating seemingly unrelated regressions and tests for aggregation bias. *Journal of the American Statistical Association, 57*, 348–368.

Zellner, A., & Theil, H. (1962). Three-stage least squares: simultaneous estimation of simultaneous equations. *Econometrica, 30*, 54–78.

Chapter 3
Partial Least Squares Path Analysis

We begin our review of modern path analysis tools with partial least squares path analysis software. PLS-PA has achieved near-cult-like stature within its circle of practitioners, but is not without its critics. Many issues arise from PLS-PA not being a proper statistical "methodology"—it has failed to accumulate a body of statistical research on assumptions, the role of data, objectives of inference, statistics, or performance metrics. Rather, PLS consists of a half dozen or so software packages that, though only lightly documented, seem to be able to conjure path estimates out of datasets that other methodologies reject as inadequate. This chapter explores whether PLS-PA software really possesses some "secret sauce" that makes it possible to generate estimates from weak data, or conversely, whether such imputed path structures may indeed be illusory.

PLS-PA lacks many of the performance and fit statistics that competing methods offer. When fit statistics do actually exist for the PLS-PA method, they tend to be loosely documented or lack formal statistical development in research papers. The development of PLS path analysis began as a legitimate search for solutions to statistical problems that presented themselves in the 1950s and 1960s. It has been superseded by better methods and most contemporary research disciplines have rejected PLS-PA software as an accepted method for data analysis, despite its practice in a few academic niches. For this reason, this chapter tries to fill the gap in statistical literature on PLS-PA while avoiding the risk of legitimizing a controversial and potentially misleading approach to data analysis.

The PLS moniker itself is purposely misleading, and has served to confound intellectual boundaries as well as terminology of the PLS culture since its inception in the early 1960s. Herman Wold developed partial least squares (PLS) regression in the 1950s out of related algorithms he had earlier applied to generating canonical correlations (i.e., correlations between pairs of groups of variables). He also applied his canonical correlation algorithms to latent variable path models; this became known as PLS, even though it did not involve partial least squares regression (PLSR), and its development was entirely separate from PLS regression. Nonetheless, it is not uncommon to see PLS articles confound the terminology of

© Springer International Publishing Switzerland 2015
J.C. Westland, *Structural Equation Models*, Studies in Systems,
Decision and Control 22, DOI 10.1007/978-3-319-16507-3_3

path modeling and PLS regression even though the two have nothing to do with each other. This may be used divisively to argue, for example, that path analysis is being used for spectral analysis, when in fact it is regression that is being used (since both are named "PLS" by the community), or that path analysis is using widely accepted regression methods. This purposeful confusion is compounded by a lack of documentation and supporting research for the software algorithms, a lack of agreement in statistics reported by competing PLS software, and obfuscation by reference to a single "PLS algorithm" (W. W. Chin, 1998, 2010b; W. W. Chin & Dibbern, 2010; W. W. Chin & Newsted, 1999).

3.1   PLS Path Analysis Software: Functions and Objectives

Wright's path analysis grew in popularity in the 1950s. Researchers in psychometrics, sociology, and education were particularly interested in fitting data to models comprising unobservable quantities such as intelligence, happiness, and effort. These "latent" variable path models could not be fit with Pearsonian correlations; they rather required more complex underlying modeling assumptions.

Wold (1961) had spent decades developing his algorithms for principal component analysis (PCA) (where his "components" could easily be identified with "latent variables") and with canonical correlation, and proposed that path coefficients be pairwise estimated (concurring with Wright's method) using his canonical correlation algorithm. His particular canonical correlation algorithm was created with the objective of maximizing the correlation between any two latent variables on a path. The overall effect on the model was to significantly overstate the correlation (i.e., the path coefficient) between any two latent variables on the path.

Wold's method was guaranteed to generate a path with significant path coefficients for every dataset, since any two variables are likely to be correlated whether or not there really exists any actual causal relationship. This makes it very easy for lazy researchers to "analyze" sloppy, poorly constructed, or misguided datasets, and yet find a path structure to support whatever theory is convenient or popular.

3.2   Path Regression

Sewall Wright's path analysis was widely used in genetics and population studies in the first part of the twentieth century. During that time, Wright attempted to interest researchers in the fields of psychometrics, survey research, and econometrics in path models, seeing similarities to problems in population studies. Wright's (1921, 1934) original path analysis defined the links between variables as correlations; causal (directional) arrows and specific restrictions on recursive paths were assumed a priori. Wright's widespread popularization of path analysis encouraged statisticians to consider other algorithms for computing path coefficients.

During a sabbatical year at the University of Chicago, his path analysis was discussed widely with econometricians who favored regression coefficients. These discussions influenced the subsequent applications of the work of a number of statisticians—in particular Herman Wold and his student Karl Jöreskog.

Econometricians in the 1950s were rapidly developing their field around regression analysis; Tukey (1954) advocated systematic replacement in path analysis of the dimensionless path coefficients by the corresponding concrete path regressions. Wright subsequently took to referring to these as the "standardized" and "concrete" forms of path analysis. The "concrete" form came to dominate computer-intensive path analysis that is used today. Wright (1960) argued convincingly that estimating the concrete form was unnecessarily complex (especially in the days before electronic computers) and that concrete estimators could be recovered from the standardized estimators anyway.

Geneticists Turner and Stevens (1959) published the seminal paper presenting the mathematics of *path regression* (a term coined by Wright). Turner and Stevens' paper ushered in modern path analysis, and their mathematics provided the basis for the computer-intensive techniques developed in the 1970s and 1980s.

During the same period, various social science fields—especially psychometrics and education—were investing significant effort in standardizing and quantifying measures of abstract concepts such as intelligence, scholastic achievement, personality traits, and numerous other unobservables that were increasingly important to US national planning and funding in the 1950s. The approach to measuring unobservable (or latent) quantities was essentially to "triangulate" them by measuring numerous related and measurable quantities. For example, intelligence tests might require an individual to spend several hours answering questions, performing tasks with speed and accuracy, problem solving, and so forth in order to assess one unobservable quantity—the intelligence quotient. These problems were more naturally suited for the canonical correlation approaches of Wold and Hotelling than they were for approaches that restricted theorists to observable variables. Wold showed that his iterative approach to computing path correlations produced the same estimates as the closed-form algebraic method of Hotelling (1936). By the late 1970s, Wold's group had implemented his canonical path correlations in Fortran computer software (J. B. Lohmöller, 1981) which generated a sequence of canonical correlations along paths on the network. This software came to be called PLS-PA.

The PLS-PA designation has caused no end of confusion in the application of Lohmöller's software, which was casually and somewhat gratuitously called "partial least squares" as a marketing ploy. Wold was well known for his development of the entirely distinct partial least squares *regression* NIPALS algorithm. NIPALS provided an alternative to OLS using a design matrix of dependent and independent variables, rather than just the independent variables of OLS. PLSR tends to work well for multicollinear data, but otherwise offers no advantage over OLS.

Controversies surround the various interpretations of coefficients. The coefficients for any given model that are generated by a particular software package are likely to diverge significantly from those computed by an alternative software package. This has created problems for interpretation and even defense of construct validity, which have been documented in numerous studies (e.g., Bollen & Ting, 2007; Henseler & Fassott, 2010; Lohmoller, 1988; McArdle, 1988; McArdle & Epstein, 1987; Pages & Tenenhaus, 2001; C. M. Ringle & Schlittgen, 2007; C. M. Ringle, Wende, & Will, 2005a; C. M. Ringle, Wende, & Will, 2005b; Tenenhaus & Vinzi, 2005; Tenenhaus, Vinzi, Chatelin, & Lauro, 2005).

3.3   Herman Wold's Contributions to Path Analysis

Herman Wold brought two important innovations to path analysis: (1) latent variables which he conceived as principal components of the indicators and (2) a widely used tool to estimate path regression coefficients (versus Wright's earlier correlation coefficients). Further, a large bit of the innovation in path models evolved around Herman Wold's work in the 1950s and 1960s. Hotelling's (1933) seminal work in PCA proposed an algorithm to compute the first principal component as a weighted linear composite of the original variables with weights chosen so that the composite accounts for the maximum variation in the original data. Wold's work in the 1940s and 1950s improved on Hotelling's computational methods. His work led eventually to regression algorithms for principal component regression (PCR) and PLSR which computed regression coefficients in situations where data was highly multicollinear (Fig. 3.1).

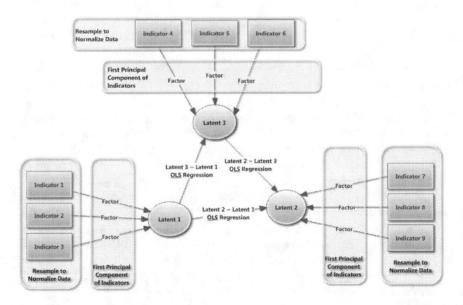

Fig. 3.1   Example of PLS path analysis model

As a by-product of this work, he started promoting the principal components as representations of abstract (latent) constructs. Latent abstractions proved useful in the evolving fields of psychometrics and sociological surveys, and were widely adopted in the 1950s and 1960s. Social scientists need to model many abstract and unobservable constructs—things like future intentions, happiness, and customer satisfaction. Though indirectly observable, there were numerous surrogates that could provide insights into such abstract (or latent) constructs—these observable surrogates were called "indicators" of the latent variable.

Wold helped this evolution along by proposing modifications to Wright's path analysis in the following fashion:

1. Let the research choose indicators for each latent construct in advance of any statistical analysis.
2. Compute the first principal component of each cluster of indicators for a specific latent variable.
3. Construct a synthetic latent variable equal to the sum of indicator value multiplied times factor weights from the first principal component for each observation.
4. Compute a coefficient between each pair of latent variables in the model using, for example, either an OLS or a PLSR regression on the first PCA components of the treatment and response (i.e., tail and head of the link arrow) latent variables. In the OLS case Wold called PCA-OLS setup a PCR. Unless the correlations between any two variables are greater than 0.95, both methods produce nearly the same coefficient.
5. Compute each regression coefficient around the model following the link arrows following Wright's three path laws.
6. Outer and inner models

   (a) The network model of latent variables is called the "structural model" or sometimes the "*inner*" model. The term "structural equation model" came about from this use, which Wold borrowed from the matrix terminology of systems of equation regression approaches developed at the Cowles Commission.
   (b) The clusters of indicators for each latent variable (with links being the factor weights on the first principal component) are sometimes called the "*outer*" model.

7. Formative and reflective links

   (a) If the researcher chooses the indicators before the latent variables, the links are called "formative" because the factor (link) weights and the latent variable are *formed* from the first component of the PCA.
   (b) If the researcher chooses the latent construct before the indicators, the links are called "reflective" because the factor (link) weights are believed to *reflect* the abstract (latent) construct. This belief must be validated by reviewing the first component of the PCA, usually through a statistic like Cronbach's alpha.

3.4   Possible Choices for Path Coefficients: Covariance, Correlation, and Regression Coefficients

Following Tukey (1954) path modeling adopted a more colorful palette of path metrics, incorporating covariances, variances, and regression coefficients, as well as correlations, which were Sewall Wright's preference for path coefficients (see Wright, 1960). Prior to surveying the detailed methods which comprise modern path analysis, it would be beneficial to recap the interpretation of each of these particular measures.

3.4.1   Covariance and Variance

Variance is the second central moment about the mean of a single variable. The square root of the variance is the standard deviation, and provides a linear measure of the variation in that variable.

   Covariance is a measure of how much two random variables change together. If the variables tend to show similar behavior, the covariance is a positive number; otherwise if the variables tend to show opposite behavior, the covariance is negative. The sign of the covariance describes the linear relationship between the variables. The magnitude of the covariance is difficult to interpret; thus correlation is typically a better statistic of magnitude of behavioral relationships.

3.4.2   Correlation

Correlation is the normalized version of the covariance, obtained by dividing covariance by the standard deviations of each of the variables. Correlation ranges from −1 to +1. Several different formulas are used to calculate correlations, but the most familiar measure is the Pearson product–moment correlation coefficient, or Pearson's correlation. Correlations are simple to interpret and to compare to each other because of their normalized range. Correlations between unobserved (latent) variables are called canonical (for continuous data) or polychoric (for ordinal data) correlation. Correlations provide useful summarizations of large datasets into single metrics, but Fig. 3.2 illustrates how misleading such extreme summarizations can become.

3.4.3   Regression Coefficients

Regression coefficients provide another measure of the way in which two random variables change together. Many differing approaches to regression are available, each with differing fit objectives, and often involving more than just two variables.

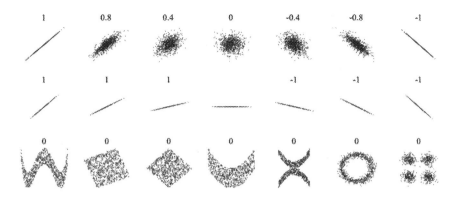

Fig. 3.2 Scatterplots and the danger of linear assumptions—pairs of random variables with Pearsonian correlations of 0, −1, and +1

Thus there is no universal definition of a regression coefficient. But generally a regression coefficient describes the relationship between a single predictor variable and the response variable—predictor and response imply a causality, with the predictor causing a predictable (based on the regression coefficient) response when all the other predictor variables in the model are held constant. A regression coefficient (often referenced with the Greek letter β) is interpreted as the expected change in the response variable (in some set of measurement units) for a one-unit change in the predictor when all of the other variables are held fixed. Regressions commonly assume linear relationships, but such assumptions may be highly misleading (as are correlations, which are also inherently linear) (Fig. 3.3).

Dijkstra (1983, p. 76) observed that Herman Wold was generally skeptical of covariance structure methods, because they too often required the assumption of normal datasets: "Wold questions the general applicability of LISREL because in many situations distributions are unknown or suspected to be far from Normally distributed. Consequently it seemed reasonable to abandon the maximum likelihood approach with its optimality aspirations and to search for a distribution-free, data-analytic approach aiming only at consistency and easy applicability."

Dijkstra (1983, p. 76) notes that after developing the PLSR algorithms, "Wold and affiliated researchers Apel, Hui, Lohmöller and S. Wold were, and at present still are, almost exclusively concerned with the development of various algorithms." Lohmöller's algorithms (J. B. Lohmöller, 1981) were the most advanced and, as we have seen, have formed the basis for modern PLS path analysis packages.

Empirical researchers begin (and may too often end) their PLS path modeling data analysis with one of several software packages available(see Table 3.1). Wold's work predated modern desktop computers, and the first widely used computer implementation of path regression only appeared in the late 1980s (Lohmoller, 1988). With the exception of semPLS, SmartPLS and SPAD-PLS, the other PLS path analysis software use Lohmöller's software code. Fortunately two thorough

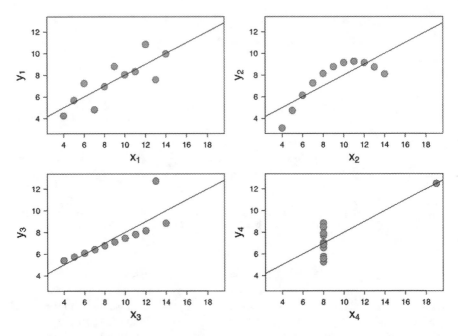

Fig. 3.3 Four sets of data with the same correlation of 0.816 and the same regression coefficient (Anscombe's quartet, Anscombe, 1973)

investigations of software methodologies (Temme & Kreis, 2006; Tenenhaus et al., 2005) of software algorithms have been published that provide insight into the internal operations of these packages.

3.5   Lohmöller's PCA-OLS Path Analysis Method

Of particular importance in the commercial software packages is Lohmöller's PCA-OLS (what Wold had termed PCR) implementation of path regression coefficient estimation. This is used as the default estimation method in all of the commercial PLS path analysis software packages, though in fact it is completely different than the PLSR estimation originally envisioned by Herman Wold. Lauro and Vinzi (2002) and Chatelin, Vinzi, and Tenenhaus (2002) provide detailed descriptions of Lohmöller's PCA-OLS methodology, though Lauro and Vinzi (2002) complain that Lohmöller's LVPLS 1.8 *"is only available in a DOS version. It presents important limitations for the number of observations and it is of a very difficult use, completely inadequate for an industrial use."* They are probably describing the motivation for the plethora of GUI wrappers for LVPLS that were developed and sold independently, and now constitute many commonly used PLS path analysis packages.

Table 3.1 Commercial PLS path analysis software products (adapted from Temme & Kreis, 2006)

PLS software	Description
LVPLS (Lohmoller, 1988)	DOS-based program (Lohmoller, 1988) with two modules for estimating paths: LVPLSC analyzes the covariance matrix, and LVPLSX uses a hybrid PCA-OLS method. Results are reported in a plain text file. The program offers blindfolding and jackknifing as resampling methods.
PLS-GUI (Li, 2005)	Windows-based graphical user interface (GUI) wrapper for Lohmöller's LVPLS code. It uses the covariance structure analysis LVPLS, and has more in common with approaches like as the first LISREL version in the early 1970s. PLS-GUI produces the same output as LVPLS.
VisualPLS (Fu, 2006a)	Another GUI wrapper for Lohmöller's LVPLS PCA-OLS method. It supports graphics in a pop-up window. Based on the graphical model, the program produces a separate LVPLS input file, which is run by LVPLSX (pls.exe). Various resampling methods are provided.
PLS-Graph (Chin, 2003)	Another GUI wrapper for Lohmöller's LVPLS PCA-OLS routine (LVPLSX). A limited set of drawing tools are provided for the path diagram, which generates a proprietary input file which cannot be processed by LVPLS. Results are provided in a reformatted LVPLS output. Various resampling methods are provided.
SPAD-PLS (Test&Go, 2006)	SPAD-PLS is part of the data analysis software SPAD offered by the French company Test&Go. Models are specified with a menu or graphically in a Java applet. This is the only available software package which actually uses PLS regression,. Transformations of latent variables (squares, cross-products) can be specified. Various resampling methods are provided.
semPLS (Monecke and Leisch, 2012)	The semPLS R-language package is the most professionally documented PLS package currently in existence, with perhaps the most complete and honest exposition of a PLS path analysis algorithm available. It also benefits from the full contingent of R tools, packages, and language for pretesting, graphics, and fitting.
SmartPLS (Ringle et al., 2005b)	The SmartPLS "drag and drop" interface is the best of the commercial PLS packages, and uses OLS and FIMIX algorithms. Various resampling methods are provided. The authors conduct an active and informative user forum at their website.

Lohmöller's PLS path analysis algorithm can be approximated with the following steps:

1. Cluster all of the indicator variables into latent variable groupings—either judgmentally (based on the intentions of the survey questions, for example) or using principal component analysis.
2. Define each latent variables as linear function of indicator variables by assigning factor loadings to each indicator-latent variable link.
3. Choose the first component of a PCA on each cluster of factors to define the factor loadings.
4. Pairwise regress each latent variables linear combination of factor weighted indicators; this is the path coefficient
5. Repeat this procedure following Wright's path diagram constraints until all paths are estimated. Continue to cycle through the model until the estimates converge.

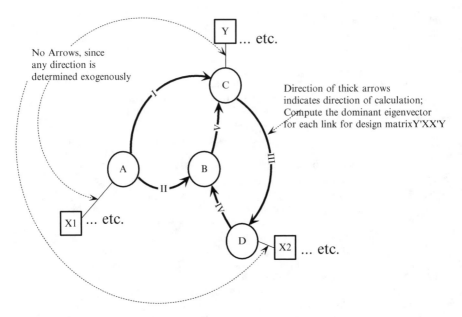

Fig. 3.4 Pairwise calculation of path coefficients with regression

Lohmöller presents these in terms of the "outer" models—the first principal component of the indicator variables assigned to a latent variable—and "inner" model—the sequence of OLS path regressions on these latent variables (see Lauro & Vinzi, 2002) (Fig. 3.4).

3.6 PLS Path Analysis vs. PLS Regression

Software vendors have created significant confusion about the methodology they use by including "PLS" in their labeling. Though Herman Wold's original intention was to implement Wright's path regression with PLSR, the implementations created by Jöreskog and Lohmöller in the 1980s did not use PLSR, rather applied two alternate approaches—(1) covariance structure modeling and (2) a hybrid principal component analysis and ordinary least squares (PCA-OLS) method. Herman Wold himself contributed to the naming confusion, attributing various terms and methods to the acronym PLS.

A brief review of PLSR and PCR is needed at this point. PLSR and PCA algorithms both estimate the values of factor loadings that help them achieve their statistical objectives. In PCA, that is the unrestricted choice of a set of latent variables (called principal components in PCA) that maximize the explanation of sample variance. PLSR is restricted to maximize the covariance between predictor and response latent variables, fundamentally assuming that the latent variables being studied are structured in a causal model—i.e., the structural equation model of unobserved variables—that has dichotomized latent variable predictors and responses.

In theory, PLSR should have an advantage over PCR. One could imagine a situation where a minor component in independent variables is highly correlated with the dependent variable; not selecting enough components would then lead to very bad predictions. In PLSR, such a component would be automatically present in the first component (latent variable). In practice, however, there is hardly any difference between the use of PLSR and PCR; in most situations, the methods achieve similar prediction accuracies, although PLSR usually needs fewer latent variables than PCR. Both behave similar to ridge regression (Frank & Friedman, 1993).

The name PLSR is itself misleading, a fact that was apparent to Herman Wold when he first introduced it in the 1960s. Wold (1966) described a procedure to compute principal component eigenvalues by an iterative sequence of OLS regressions, where loadings were identical to closed-form algebraic methods. This was the origin of the term partial least squares, to describe the iterative OLS regressions used to calculate principal components. In his approach the eigenvalues were interpreted as the proportion of variance accounted for by the correlation between the respective "canonical variates" (i.e., the factor loadings) for weighted sum scores of the two sets of variables. These canonical correlations measured the simultaneous relationship between the two sets of variables, where as many eigenvalues are computed as there are canonical roots (i.e., as many as the minimum number of variables in either of the two sets). Wold showed that his iterative approach produced the same estimates as closed-form algebraic method of computing the (Hotelling, 1936) cross-covariance matrices in canonical correlation analysis.

The name partial least squares *regression* itself created confusion, and Wold tried in the 1970s to drop regression from the name, without success. As a consequence, at various times, suggestions have arisen to compute regression goodness-of-fit statistics for PLS path analysis, such as R-squared, F-statistics, and t-statistics; indeed, PLS path analysis computer packages may sometimes even report such fit measures. These numbers exist for individual paths, but are meaningless for the model as a whole.

One particular software package described in the previous section, SPAD-PLS (Test&Go, 2006), did take Wold at his word, and does apply PLSR (as an option) to computing path regression coefficients. Thus it is worthwhile at this juncture to elaborate on the methods developed by Wold and justify why these might be desirable in a path model.

In well-controlled surveys (McArdle & Hamagami, 1996), OLS may be substituted for PLSR. This fact has not been lost on developers of PLS path analysis software. Several studies (Mevik & Wehrens, 2007; Temme & Kreis, 2006; Tenenhaus et al. 2005) reviewed the algorithms used in existing PLS path analysis software, and investigated estimation performance. A variety of estimation procedures—including OLS and PLSR—were applied across the packages; Temme and Kreis (2006) note that "SPAD-PLS is the only program which takes this problem into account by offering an option to use PLSR in the estimation." The general inclination is to apply OLS to the paths, and the factor loadings, path coefficients, and R^2 are essentially what one would get by regressing the first principal component—a one latent variable principal component regression (Table 3.2).

Table 3.2 Strengths and weaknesses of the PLS analysis

Strengths	Weaknesses
PLS path analysis is able to model multiple dependents as well as multiple independent variables	PLS path analysis software generally uses OLS regression methods (the "PLS" in the name is a misnomer initiated by Wold) and thus provides no better estimates than traditional sequential limited information path analysis.
PLS path analysis software produces estimates from small datasets (though heavy use of sampling makes these estimates difficult to interpret or verify)	The small-sample properties of PLS path analysis software are not inherent in the regression algorithm, rather result from intensive and often poorly modeled and justified resampling of sample data.
PLS predictions are able to handle multicollinearity among the independents	PLS is less than satisfactory as an explanatory technique because it is low in power to filter out variables of minor causal importance. PLS estimator distributional properties are not known; thus the researcher cannot assess "fit," and indeed the term probably is not meaningful in the PLS context.
Because PLS path analysis estimates one path at a time, models do not need to be reduced and identified, as in systems of equation regression models	PLS path analysis is often used to process Likert scale data—which may be considered either ordinal or polytomous Rasch data. Heavy use of Gaussian resampling is used to force estimator convergence in the software algorithm, which makes assessment of the validity of coefficients difficult.
Heavy reliance on resampling allows PLS path analysis to compute estimates in the face of data noise and missing data	Theory-driven introduction of prior information into the resampling distribution models and testing is questionable, because the "data are not able to speak for themselves."
In contrast to LISREL SEM. A model is said to be identified if all unknown parameters are identified.	PLS estimator distributional properties are not known; thus the researcher cannot assess "fit," and indeed the term probably is not meaningful in the PLS context.

3.7 Resampling

All of the PLS path analysis software packages have touted their ability to calculate coefficients from datasets that would be too small for covariance structure modeling methods, or other commonly used statistical methods. Unfortunately, this claim is somewhat misleading, as the software accomplishes this through computer-intensive resampling—estimating the *precision* of sample statistics (medians, variances, percentiles) by replicating available data (jackknifing) or drawing randomly with replacement from a set of data points (bootstrapping). Such techniques are steeped in controversy, and it is not my intention to wade into the debate.

Resampling should generally be avoided for another reason. Assume that statistical precision is another way of specifying the Fisher information in the sample. Then a dataset that is resampled from a given set of observations has exactly the same Fisher information as the original dataset of observations. The only way that the reported statistics will improve is if the researcher (erroneously) assumes that the additional observations are *not* resampled, but are new, independent observations. This is not completely honest—the only honest way to collect more information is to go out and increase the sample size.

3.8 Measures

PLS path analysis does not "fit" the entire model to a dataset in the sense of solving an optimization subject to a loss function. A PLS path regression is simply a disjoint sequence of canonical correlations of user-defined clusters of indicator variables. As a consequence, fit measures of the whole structural model are not meaningful— the only fit statistics or performance measures possible are ones that apply to individual links. Even the interpretation of standard OLS regression fit statistics on individual links, such as R-squared, F-statistics, or t-statistics, is confounded by the synthetic nature of the latent variables.

Despite this, numerous ad hoc fit measures have been proposed for PLS path analysis, and many if not most of these are misleading (for example, see Chin, 2010a, 2010b; Chin & Dibbern, 2009, 2010; Götz, Liehr-Gobbers, & Krafft, 2010; Henseler & Fassott, 2010; Henseler, Ringle, & Sinkovics, 2009; Hulland, Ryan, & Rayner, 2010; Marcoulides, Chin, & Saunders, 2009; Ringle, Wende, & Will, 2010; Sarstedt, 2008; Vinzi, Chin, & Henseler, 2009; Wold, 1973, for examples of various ad hoc metrics with discussion). Some of these "fit" measures have arisen from a confusion about what PLS actually does. Wold (1973, 1974) finessed the role of PLS with explanations of "plausible causality"—a hybrid of model specification and exploratory data analysis—which added to rather than resolving confusion over his innovation.

3.9 "Limited Information"

PLS path analysis is often called a "limited information" method without bothering to further define what exactly this means. But this is precisely where the lack of fit statistics generates the greatest impact on PLS path analysis.

Limited information in path analysis implies that OLS estimators on individual pairwise paths will, in most practical circumstances, replicate the results of PLS path analysis software. But in specific ways, separate regressions could improve on PLS path model software programs. With individual regressions, R-squared, F-statistics, and residual analysis through graphing and other tests can reveal a great deal of information about the underlying data. PLS path analysis software too often obscures such information in the data by (1) resampling, which imposes prior beliefs about the population on the data, and (2) overreaching by claiming to assess the entire model at once. Research is generally better served by a full information method such as covariance approaches (e.g., LISREL, AMOS) or a system of equations approach.

3.10   Sample Size in PLS-PA

A stubborn mythology surrounds notions of data adequacy and sample size in PLS-PA that originated with an almost offhand comment in Nunnally, Bernstein, and Berge (1967) who suggested (without providing supporting evidence) that in estimation "a good rule is to have at least ten times as many subjects as variables." Justifications for this *rule of 10* appear in many frequently cited publications (Barclay, Higgins, & Thompson, 1995; W. W. Chin, 1998, 2010a, 2010b; W. W. Chin & Dibbern, 2009, 2010; W. W. Chin & Newsted, 1999; Kahai & Cooper, 2003). But Goodhue, Lewis, and Thompson (2006); Boomsma (1982, 1985, 1987); Boomsma and Hoogland (2001); Ding, Velicer, and Harlow (1995); and others have studied the *rule of 10* using Monte Carlo simulation, showing that the rule of 10 cannot be supported. Indeed, sample size studies have generally failed to support the claim that PLS-PA demands significantly smaller sample sizes for testing than other SEM methods.

Adaptive bias becomes a significant problem as models grow more complex. Rather than seeking absolute truth or even rational choice, our brains adapt to make decisions on incomplete and uncertain information input, with an objective of maximum speed at minimum energy expenditure (Gilbert, 1998; Henrich & McElreath, 2003; Neuhoff, 2001). This manifests itself in anchoring and adjustment on preconceived models (Kahneman & Tversky, 1979) and stronger and stronger biases towards false positives as models grow more complex. The latter problems might be considered one of intellectual laziness, as the brain does not want to expend effort on thinking about an exponentially increasing number of alternative models.

Statisticians use the term "power" of a test to describe the ability of a method to test a particular model and dataset to minimize false positives. Complex network models drive the power of tests towards zero very quickly. Where multiple hypotheses are under consideration, the powers associated with the different hypotheses will differ as well. Other things being equal, more complex models will inflate both the type I and type II errors significantly (Bland & Altman, 1995).

Nearly every study investigating Nunnally's "*rule of 10*" for sample size has found it to yield sample sizes that are many orders of magnitude too small. Studies relying on these small samples are unreliable with significantly inflated type I and II error rates. Four early studies, Bollen (1989), Bollen and Ting (2007), Bentler (1990), and Bentler and Mooijaart (1989), rejected the rule of 10 as fiction, and suggested a possible 5:1 ratio of sample size to number of free parameters. Monte Carlo studies conducted in the 1980s and 1990s showed that SEM estimator performance is not linearly or quadratically correlated with the number of parameters (Browne & Cudeck, 1989, 1992, 1993; Browne, Cudeck, Tateneni, & Mels, 2002; Gerbing & Anderson, 1988; Geweke & Singleton, 1981). Going further Velicer, Prochaska, Fava, Norman, and Redding (1998) reviewed a variety of such recommendations in the literature, concluding that there was no support for rules positing a minimum sample size as a function of indicators. They showed that for a given sample size, a convergence to proper solutions and goodness of fit were favorably influenced by

(1) a greater number of indicators per latent variable and (2) a greater saturation (higher factor loadings). Several studies (Marsh & Bailey, 1991; Marsh, Byrne, & Craven, 1992; Marsh, Wen, & Hau, 2004; Marsh & Yeung, 1997, 1998) concluded that the *ratio of indicators to latent variables* rather than just the number of indicators is a substantially better basis on which to calculate sample size, reiterating conclusions reached by Boomsma (1982, 1985, 1987) and Boomsma and Hoogland (2001). We will revisit this problem later, and provide criteria for control of error inflation and adaptive bias in sample selection for structural equation models.

The availability of PLS-PA software packages allows a relatively straightforward Monte Carlo exploration of statistical power and sample size. Fortunately Monecke and Leisch (2012) have provided an accessible implementation on the R-language platform (*semPLS*) that can be used to explore its otherwise arcane characteristics. As this section will show, there are curious idiosyncrasies of PLS-PA that set it apart from other widely used statistical approaches. We can show that (1) contrary to the received mythology, PLS-PA path estimates are biased and highly dispersed with small samples; (2) sample sizes must grow very large to control this bias and dispersion with dispersion $\propto \dfrac{1}{\log(\text{sample size})}$ and bias $\propto \dfrac{1}{\text{sample size}}$; and (3) the power and significance of PLS-PA hypothesis test are roughly the same as for 2SLS models, concurring with Dhrymes (1972, 1974), Dhrymes and Erlat (1972), and Dhrymes et al. (1972), and for small samples are low at most effect levels, yielding an excessive number of false positives.

Various studies (C. M. Ringle & Schlittgen, 2007; C. M. Ringle et al., 2005a; C. M. Ringle et al., 2010; C. M. Ringle et al., 2005b; C. M Ringle, Sarstedt, & Straub, 2012) have reviewed the use of the (J. B. Lohmöller, 1981; Lohmoller, 1988; J. -B. Lohmöller, 1989) algorithm methods and estimators. Monecke and Leisch (2012) noted that all of the PLS path model software use the same Lohmöller (1989) algorithm on an ad hoc iterative estimation technique. C. M. Ringle, Sarstedt, and Straub (2012) surveyed a subset of studies using PLS-PA for testing, noting that three-quarters of studies justify the application of PLS path analysis software in a single paragraph at the beginning of the data analysis citing PLS' use of either small sample sizes (36.92 %) or non-normal data (33.85 %); additionally, more than two-thirds of studies violate the standard assumptions made in estimation with PLS-PA.

Monecke and Leisch (2012) describe their implementation of Lohmöller's algorithm, as follows. Assume a given path model, for example $A \rightarrow B \rightarrow C$. Assume that $\{A,B,C\}$ are latent (i.e., unobservable) variables, and that each comprises a pair of observable "indicator" variables: $A = \sum_i \alpha_i A_i;\ B = \sum_i \beta_i B_i;\ C = \sum_i \gamma_i C_i; i = 1, 2$. The PLS software maximizes a scaled canonical correlation between pairs of variables by iteratively stepping through the model and adjusting factor weights $\{\alpha_1,\alpha_2,\beta_1,\beta_2,\gamma_1,\gamma_2\}$. First, on path $A \rightarrow B$, weights $\{\alpha_1,\alpha_2\}$ are set to initial values, say $\{\alpha_1, \alpha_2\} = \{0.5, 0.5\}$, and weights $\{\beta_1,\beta_2\}$ are manipulated to maximize the path coefficient (i.e., a scaled canonical correlation) on $A \rightarrow B$. The process is repeated for path

$B \to C$ keeping the computed weights $\{\beta_1, \beta_2\}$ and choosing $\{\gamma_1, \gamma_2\}$ to maximize the path coefficient on $B \to C$. This is repeated in a cycle $C \to A \to B \to C \to A \to B \to \ldots$ until the change in path coefficient values is less than some preset value. The fitting of data to the path model only occurs on individual pairs of latent variables—it is piecewise. Researchers call this "limited information" fitting of the data, meaning that all of the data outside of a particular path is ignored in maximizing a path coefficient. "Limited information" approaches to path modeling generate highly inflated path coefficients, and many researchers like that as it lowers their workload in proving a theory. When model paths are preselected (for example $A \to B$ and $B \to C$ are preselected out of three possible paths $C \to A$, $A \to B$ and $B \to C$) these "limited information" approaches have significantly higher probability of statistically confirming the preselected path $A \to B \to C$ than would a method that considered all of the data simultaneously. This is likely why its supporters assume that PLS-PA works well with small sample size and non-normal data that would not yield statistically significant results using standard approaches.

To analyze the behavior of PLS-PA algorithms, an inner (latent variable) model was tested with the R-code *semPLS* package on data that follows a zero-centered normal distribution, but is expressed as indicator variables that are Likert-scaled 1:5 integer values. This is a standard path setup that is used in many social science research papers. Each of the latent variables was measured with three indicator variables that are Likert-scaled 1:5 integer values. Every indicator is statistically independent of any others, and the population correlation structure is an 18×18 identity matrix (complete independence of observed variables). Random data for this study was generated using the Mersenne-Twister algorithm (Matsumoto & Nishimura, 1998) which is a twisted GFSR with period $2^{19937} - 1$ and equidistribution in 623 consecutive dimensions (over the whole period). The random "seed" is a 624-dimensional set of 32-bit integers plus a current position in that set. The following figure shows the R-code used for estimator tests (Figs. 3.5 and 3.6).

The parameter estimates for each path over 20 trials with a sample of 20 observations. Estimator values span the entire range of possible values $\{-1,1\}$. If the PLS-PA algorithm accurately estimated path coefficients from this population, these values would be zero (since all the observations are independent) (Fig. 3.7).

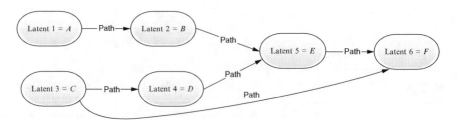

Fig. 3.5 Path model analyzed with "semPLS" package

```
## Specify the model from two .csv files, per the semPLS documentation
sm = as.matrix(read.csv("sm.csv"))
mm = as.matrix(read.csv("mm.csv"))

repeat {        ## generate coefficient tables until you push the 'stop' button

## Create a data.frame of manifest variables, filled with Likert 1:5 integer
values generated by a Mersenne-Twister algorithm
size = 200    ## Sample size of uniform random Likert values 1:5
plsdata=data.frame(A1 = floor(1+5*runif(size)), A2 = floor(1+5*runif(size)),A3
= floor(1+5*runif(size)),B1 = floor(1+5*runif(size)),B2 =
floor(1+5*runif(size)),B3 = floor(1+5*runif(size)),C1 =
floor(1+5*runif(size)),C2 = floor(1+5*runif(size)),C3 =
floor(1+5*runif(size)),D1 = floor(1+5*runif(size)),D2 =
floor(1+5*runif(size)),D3 = floor(1+5*runif(size)),E1 =
floor(1+5*runif(size)),E2 = floor(1+5*runif(size)),E3 =
floor(1+5*runif(size)),F1 = floor(1+5*runif(size)),F2 =
floor(1+5*runif(size)),F3 = floor(1+5*runif(size)))

## Estimate path coefficients
library("semPLS")
plsmod <- plsm(data = plsdata, strucmod = sm, measuremod = mm)
plscoeff <- sempls(model = plsmod, data = plsdata, wscheme = "centroid",
verbose=FALSE,  maxit=2000)

## Print a table of estimated path coefficients

pathtable=pathCoeff(plscoeff)
print(pathtable)
    }
```

Fig. 3.6 R-code for tests

The figure plots the path estimator standard deviation over 20 trials with sample sizes of $\{10^1, 10^2, 10^3, 10^4, 10^5, 10^6\}$ and plots these on a log-log graph. Figure 3.8 shows that on all paths, standard deviation (vertical axis) tends to scale with sample size (horizontal axis) and the path coefficient standard deviation $\propto \dfrac{1}{\log(\text{sample size})}$.

A log-linear graph (Fig 3.8) of path estimator bias shows that direction of bias on a given segment of the model path is consistent for larger and larger sample sizes. We can state approximately path coefficient bias $\propto \dfrac{1}{\text{sample size}}$. If the PLS algorithm accurately estimated path coefficients from this population, these values would be zero (since all the observations are independent) (Fig. 3.9).

For any given model, path coefficients tend to retain a particular direction of estimation bias that varies with the position of the path in the path model. In the particular model tested here, estimators towards the right side of the model (D–E, E–F, C–F) show considerably less bias than those paths to the left (A–C, C–D, B–E). This is most likely an artifact of the particular sequence of pairwise estimation chosen by the (Monecke & Leisch, 2012) implementation of the algorithm.

This brief example provides us with clear guides on how to interpret the assertions that conflate PLS-PA's ability to generate coefficients without abnormally terminating as equivalent to estimating with small samples. The various software implementations

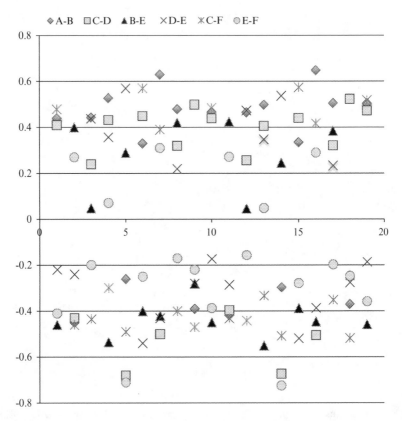

Fig. 3.7 Path coefficients over 20 trials

of PLS-PA do generate purported "goodness-of-fit" metrics such as R^2 and F-stats, but any fit statistics can only make sense in the limited context of individual path estimates. In the context of a complete path model, these statistics are meaningless. Indeed, Wold (1982) alluded to the incompleteness of his approach, referring to it as a "limited information" approach where path coefficients suggest only "plausible causalities." We can glean the following insights from this brief Monte Carlo study:

1. PLS path estimates are biased and highly dispersed when computed from small samples. In general path coefficient standard deviation $\propto \dfrac{1}{\log(\text{sample size})}$ and path coefficient bias $\propto \dfrac{1}{\text{sample size}}$. These relationships were computed against a baseline set by the Mersenne-Twister random number-generating algorithm that assures zero population correlation between indicators and thus latent variables. If the PLS algorithm accurately estimated path coefficients from this population, all path coefficients would be zero, since all of the observations are independent.

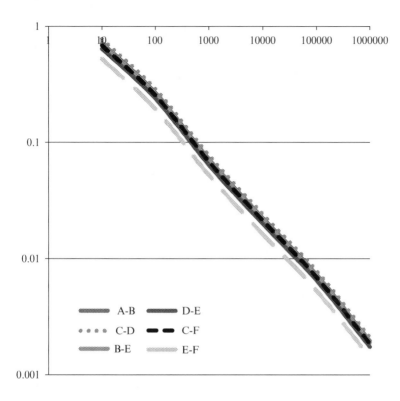

Fig. 3.8 Path coefficient standard deviation $\propto \dfrac{1}{\log(\text{sample size})}$

2. Path coefficient bias and dispersion are significant for the commonly used 5-point Likert scale data. Sample sizes must grow more rapidly than they will in regression for estimating the same data and path models, because, in contrast to PLS-PA, regression analysis has methods for incorporating distributional information (e.g., via general linear models) and analysis of residuals. Contrary to the widely cited justifications for application of PLS path analysis software, small sample sizes and non-normal data, PLS algorithms actually require significantly larger sample sizes to extract information from the population. In effect, the algorithm is throwing away useful data from the sample, but researchers fail to see this, because the software is usually able to compute some number for the path coefficient. Prior literature has conflated PLS software's ability to generate coefficients without abnormally terminating as equivalent to estimating with small samples and non-normal distributions. But this is in fact a flaw in the PLS path analysis software that allows it to generate incorrect results without generating a corresponding warning to the researcher.

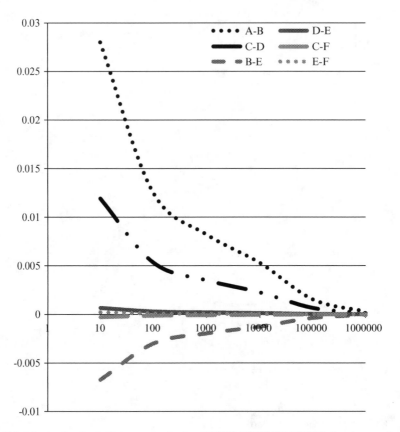

Fig. 3.9 Means of path coefficients sample sizes of $\{10^1, 10^2, 10^3, 10^4, 10^5, 10^6\}$ showing consistent positive or negative bias on particular paths

3. The power of PLS hypothesis tests is low across effect levels, leading PLS software to generate a disproportionate number of false positives. Even ignoring the
   cherry-picking of data cited by Ioannidis (2005), excessive generation of false
   positives supports the widespread publication of erroneous conclusions.

These findings contradict the widely cited justifications for application of
PLS path analysis software: either small sample sizes or non-normal data.
Responsible design of software would stop calculation when the information in
the data is insufficient to generate meaningful results, thus limiting the potential for publication of false conclusions. Unfortunately, much of the methodological literature associated with PLS software has conflated its ability to
generate coefficients without abnormally terminating as equivalent to extract-
ing information.

References

Anscombe, F. J. (1973). Graphs in statistical analysis. *The American Statistician, 27*(1), 17–21.

Barclay, D., Higgins, C., & Thompson, R. (1995). The partial least squares (PLS) approach to causal modeling: personal computer adoption and use as an illustration. *Technology Studies, 2*(2), 285–309.

Bentler, P. M. (1990). Fit indexes, Lagrange multipliers, constraint changes and incomplete data in structural models. *Multivariate Behavioral Research, 25*(2), 163–172.

Bentler, P. M., & Mooijaart, A. B. (1989). Choice of structural model via parsimony: a rationale based on precision. *Psychological Bulletin, 106*(2), 315.

Bland, J. M., & Altman, D. G. (1995). Multiple significance tests: the Bonferroni method. *BMJ, 310*(6973), 170.

Bollen, K. A. (1989). *Structural equations with latent variables* (Vol. 8). New York, NY: Wiley.

Bollen, K. A., & Ting, K. (2007). *Confirmatory tetrad analysis*. Cambridge, MA: Blackwell.

Boomsma, A. (1982). The robustness of LISREL against small sample sizes in factor analysis models. In K. G. Joreskog & H. Wold (Eds.), *Systems under indirect observation: causality, structure, prediction* (Vol. 1, pp. 149–173). Amsterdam: North-Holland.

Boomsma, A. (1985). Nonconvergence, improper solutions, and starting values in LISREL maximum likelihood estimation. *Psychometrika, 50*(2), 229–242.

Boomsma, A. (1987). *The robustness of maximum likelihood estimation in structural equation models*. Cambridge: Cambridge University Press.

Boomsma, A., & Hoogland, J. J. (2001). The robustness of LISREL modeling revisited. In R. Cudeck, S. du Toit, & D. Sorbom (Eds.), *Structural equation models: present and future. A festschrift in honor of Karl Jöreskog* (pp. 139–168). Lincolnwood, IL: Scientific Software International.

Browne, M. W., & Cudeck, R. (1989). Single sample cross-validation indices for covariance structures. *Multivariate Behavioral Research, 24*(4), 445–455.

Browne, M. W., & Cudeck, R. (1992). Alternative ways of assessing model fit. *Sociological Methods & Research, 21*(2), 230–258.

Browne, M. W., & Cudeck, R. (1993). Alternative ways of assessing model fit. In *Sage focus editions* (Vol. 154, p. 136). Thousand Oaks, CA: Sage.

Browne, M. W., Cudeck, R., Tateneni, K., & Mels, G. (2002). CEFA: comprehensive exploratory factor analysis. Computer.

Chatelin, Y. M., Vinzi, V. E., & Tenenhaus, M. (2002). State-of-art on PLS Path Modeling through the available software.

Chin, W. W. (1998). Commentary: issues and opinion on structural equation modeling. *MIS Quarterly, 22*, 7.

Chin, W. W., Marcolin, B. L., & Newsted, P. R. (2003). A partial least squares latent variable modeling approach for measuring interaction effects: Results from a Monte Carlo simulation study and an electronicmail emotion/adoption study. *Information Systems Research, 14*(2), 189–217.

Chin, W. W. (2010a). Bootstrap cross-validation indices for PLS path model assessment. In V. E. Vinzi, W. W. Chin, J. Henseler, & H. Wang (Eds.), *Handbook of partial least squares* (pp. 83–97). New York, NY: Springer.

Chin, W. W. (2010b). How to write up and report PLS analyses. In V. E. Vinzi, W. W. Chin, J. Henseler, & H. Wang (Eds.), *Handbook of partial least squares* (pp. 655–690). New York, NY: Springer.

Chin, W. W., & Dibbern, J. (2009). A permutation based procedure for multi-group PLS analysis: results of tests of differences on simulated data and a cross of information system services between Germany and the USA. In V. E. Vinzi, W. W. Chin, J. Henseler, & H. Wang (Eds.), *Handbook of partial least squares: concepts, methods and applications in marketing and related fields*. Berlin: Springer.

Chin, W. W., & Dibbern, J. (2010). An introduction to a permutation based procedure for multi-group PLS analysis: results of tests of differences on simulated data and a cross cultural analysis of the sourcing of information system services between Germany and the USA. In V. E. Vinzi, W. W. Chin, J. Henseler, & H. Wang (Eds.), *Handbook of partial least squares* (pp. 171–193). New York, NY: Springer.

Chin, W. W., & Newsted, P. R. (1999). Structural equation modeling analysis with small samples using partial least squares. In R. Hoyle (Ed.), *Statistical strategies for small sample research* (Vol. 2, pp. 307–342). Thousand oaks, CA: Sage.

Dhrymes, P. J. (1972). *Distributed lags: a survey*. Los Angeles, CA: UCLA Department of Economics.

Dhrymes, P. J. (1974). *Econometrics*. New York, NY: Springer.

Dhrymes, P. J., & Erlat, H. (1972). *Asymptotic properties of full information estimators in dynamic autoregressive simultaneous models*. Los Angeles, CA: UCLA Department of Economics.

Dhrymes, P. J., Howrey, E. P., Hymans, S. H., Kmenta, J., Leamer, E. E., Quandt, R. E., ... Zarnowitz, V. (1972). Criteria for evaluation of econometric models. *Annals of Economic and Social Measurement, 1(3)*, 291–325, NBER.

Dijkstra, T. (1983). Some comments on maximum likelihood and partial least squares methods. *Journal of Econometrics, 22*(1), 67–90.

Ding, L., Velicer, W. F., & Harlow, L. L. (1995). Effects of estimation methods, number of indicators per factor, and improper solutions on structural equation modeling fit indices. *Structural Equation Modeling, 2*(2), 119–143.

Frank, I. E., & Friedman, J. H. (1993). A statistical view of some chemometrics regression tools. *Technometrics, 35*, 109–135.

Fu, J.-R. (2006a). VisualPLS—Partial least square (PLS) regression—An enhanced GUI for Lvpls (PLS 1.8 PC) Version 1.04. National Kaohsiung University of Applied Sciences, Taiwan, ROC.

Fu, J.-R. (2006b). VisualPLS—Partial least square (PLS) regression—An enhanced GUI for Lvpls (PLS 1.8 PC) Version 1.04. Retrieved from http://www2.kuas.edu.tw/prof/fred/vpls/index.html.

Gerbing, D. W., & Anderson, J. C. (1988). An updated paradigm for scale development incorporating unidimensionality and its assessment. *Journal of Marketing Research, 25*, 186–192.

Geweke, J. F., & Singleton, K. J. (1981). Maximum likelihood "confirmatory" factor analysis of economic time series. *International Economic Review, 22*(1), 37–54.

Gilbert, P. (1998). The evolved basis and adaptive functions of cognitive distortions. *British Journal of Medical Psychology, 71*(4), 447–463.

Goodhue, D., Lewis, W., & Thompson, R. (2006). *PLS, small sample size, and statistical power in MIS research*. Washington, DC: IEEE Computer Society.

Goodhue, D. L., Lewis, W., & Thompson, R. (2012). Does PLS have advantages for small sample size or non-normal data? *MIS Quarterly, 36*(3), 891–1001.

Götz, O., & Liehr-Gobbers, K. (2004). Analyse von Strukturgleichungsmodellen mit Hilfe der partial-leastsquares (PLS)-Methode. *Die Betriebswirtschaft, 64*(6), 714–738.

Hair, J. F., Ringle, C. M., & Sarstedt, M. (2011). PLS-SEM: Indeed a silver bullet. *The Journal of Marketing Theory and Practice, 19*(2), 139–152.

Henrich, J., & McElreath, R. (2003). The evolution of cultural evolution. *Evolutionary Anthropology, 12*(3), 123–135.

Henseler, Jorg, Christian M. Ringle, & Rudolf R. Sinkovics. (2009). The use of PLS in international marketing.

Henseler, J., & Fassott, G. (2010). Testing moderating effects in PLS path models: an illustration of available procedures. In V. E. Vinzi, W. W. Chin, J. Henseler, & H. Wang (Eds.), *Handbook of partial least squares* (pp. 713–735). New York, NY: Springer.

Hotelling, H. (1933). Analysis of a complex of statistical variables into principal components. *Journal of Educational Psychology, 24*(6), 417.

Hotelling, H. (1936). Relations between two sets of variates. *Biometrika, 28*(3/4), 321–377.

Hulland, John, Michael J. Ryan, & Robert K. Rayner. (2010). Modeling customer satisfaction: A comparative performance evaluation of covariance structure analysis versus partial least squares. Handbook of partial least squares (pp. 307–325). Berlin/Heidelberg: Springer.

Ioannidis, J. P. A. (2005). Why most published research findings are false. *PLoS Medicine, 2*(8), e124.

Kahai, S. S., & Cooper, R. B. (2003). Exploring the core concepts of media richness theory: the impact of cue multiplicity and feedback immediacy on decision quality. *Journal of Management Information Systems, 20*(1), 263–299.

Kahneman, D., & Tversky, A. (1979). Prospect theory: an analysis of decision under risk. *Econometrica, 47*, 263–291.

Lauro, C., & Vinzi, V. E. (2002). *Some contributions to PLS path modeling and a system for the European customer satisfaction.* Milano: Universita di Milano Bicocca. atti della XL1 riunione scientifica SIS.

Li, Y. (2005). PLS-GUI—Graphic User Interface for Partial Least Squares (PLS-PC 1.8)—Version 2.0.1 beta. University of South Carolina, Columbia, SC.

Lohmöller, J. B. (1981). *Pfadmodelle mit latenten variablen: LVPLSC ist eine leistungsfähige alternative zu LIDREL.* München: Hochsch. d. Bundeswehr, Fachbereich Pädagogik.

Lohmoller, J.-B. (1988). The PLS program system: latent variables path analysis with partial least squares estimation. *Multivariate Behavioral Research, 23*(1), 125–127.

Lohmöller, J.-B. (1989). *Latent variable path modeling with partial least squares.* Heidelberg: Physica-Verlag.

Marsh, H. W., & Bailey, M. (1991). Confirmatory factor analyses of multitrait-multimethod data: a comparison of alternative models. *Applied Psychological Measurement, 15*(1), 47.

Marsh, H. W., Byrne, B. M., & Craven, R. (1992). Overcoming problems in confirmatory factor analyses of MTMM data: the correlated uniqueness model and factorial invariance. *Multivariate Behavioral Research, 27*(4), 489–507.

Marsh, H. W., Wen, Z., & Hau, K. T. (2004). Structural equation models of latent interactions: evaluation of alternative estimation strategies and indicator construction. *Psychological Methods, 9*(3), 275.

Marsh, H. W., & Yeung, A. S. (1997). Causal effects of academic self-concept on academic achievement: structural equation models of longitudinal data. *Journal of Educational Psychology, 89*(1), 41.

Marsh, H. W., & Yeung, A. S. (1998). Longitudinal structural equation models of academic self-concept and achievement: gender differences in the development of math and English constructs. *American Educational Research Journal, 35*(4), 705.

Marcoulides, G. A., Chin, W. W., & Saunders, C. (2009). A critical look at partial least squares modeling. *MIS Quarterly, 36*, 171–175.

Matsumoto, M., & Nishimura, T. (1998). Mersenne twister: a 623-dimensionally equidistributed uniform pseudo-random number generator. *ACM Transactions on Modeling and Computer Simulation, 8*(1), 3–30.

McArdle, J. J. (1988). Dynamic but structural equation modeling of repeated measures data. In J. R. Nesselroade & R. B. Cattell (Eds.), *Handbook of multivariate experimental psychology* (pp. 561–614). New York, NY: Springer.

McArdle, J. J., & Epstein, D. (1987). Latent growth curves within developmental structural equation models. *Child Development, 58*, 110–133.

McArdle, J. J., & Hamagami, F. (1996). Multilevel models from a multiple group structural equation perspective. In G. A. Marcoulides & R. E. Schumacker (Eds.), *Advanced structural equation modeling: issues and techniques* (pp. 89–124). Mahwah, NJ: Lawrence Erlbaum.

Mevik, B. H., & Wehrens, R. (2007). The PLS package: principal component and partial least squares regression in R. *Journal of Statistical Software, 18*(2), 1–24.

Monecke, A., & Leisch, F. (2012). semPLS: structural equation modeling using partial least squares. *Journal of Statistical Software, 48*(3), 1–32. http://www.jstatsoft.org/.

Neuhoff, J. G. (2001). An adaptive bias in the perception of looming auditory motion. *Ecological Psychology, 13*(2), 87–110.

Nunnally, J. C., Bernstein, I. H., & ten Berge, J. M. F. (1967). *Psychometric theory* (Vol. 226). New York, NY: McGraw-Hill.

Pages, J., & Tenenhaus, M. (2001). Multiple factor analysis combined with PLS path modelling. Application to the analysis of relationships between physicochemical variables, sensory profiles and hedonic judgements. *Chemometrics and Intelligent Laboratory Systems, 58*(2), 261–273.

Ringle, C. M., Sarstedt, M., & Straub, D. W. (2012). Editor's comments: a critical look at the use of PLS-SEM in MIS quarterly. *MIS Quarterly, 36*(1), iii–xiv.

Ringle, C. M., & Schlittgen, R. (2007). A genetic algorithm segmentation approach for uncovering and separating groups of data in PLS path modeling. *PLS, 7*, 75–78.

Ringle, C. M., Wende, S., & Will, A. (2005). Customer segmentation with FIMIX-PLS. PLS and Related Methods – Proceedings of the PLS, vol. 5, pp. 507–514.

Ringle, C. M., Wende, S., & Will, S. (2005). SmartPLS 2.0 (M3) Beta, Hamburg, Retrieved from: http://www.smartpls.de

Ringle, C. M., Wende, S., & Will, A. (2010). Finite mixture partial least squares analysis: methodology and numerical examples. In V. E. Vinzi, W. W. Chin, J. Henseler, & H. Wang (Eds.), *Handbook of partial least squares* (pp. 195–218). New York, NY: Springer.

Sarstedt, M. (2008). A review of recent approaches for capturing heterogeneity in partial least squares path modelling. *Journal of Modelling in Management, 3*(2), 140–161.

Temme, D., & Kreis, H. (2006). PLS path modeling – a software review. *Computational Statistics & Data Analysis, 48*(1), 159–205.

Tenenhaus, M., & Vinzi, V. E. (2005). PLS regression, PLS path modeling and generalized Procrustean analysis: a combined approach for multiblock analysis. *Journal of Chemometrics, 19*(3), 145–153.

Tenenhaus, M., Vinzi, V. E., Chatelin, Y. M., & Lauro, C. (2005). PLS path modeling. *Computational Statistics & Data Analysis, 48*(1), 159–205.

Test&Go. (2006). Spad Version 6.0.0. Paris, France.

Tukey, J. W. (1954). Causation, regression, and path analysis. In O. Kempthorne (Ed.), *Statistics and mathematics in biology* (pp. 35–66). Ames, IA: Iowa State College Press.

Turner, M. E., & Stevens, C. D. (1959). The regression analysis of causal paths. *Biometrics, 15*(2), 236–258.

Velicer, W. F., Prochaska, J. O., Fava, J. L., Norman, G. J., & Redding, C. A. (1998). Detailed overview of the transtheoretical model. *Homeostasis, 38*, 216–233.

Vinzi, V. E., et al. (2009). *Handbook of partial least squares: Concepts, methods and applications in marketing and related areas. Handbooks of Computational Statistics*. Berlin/Heidelberg, Germany: Springer-Verlag.

Wold, Herman O. A. (1961). Unbiased predictors. Proceedings of the fourth Berkeley symposium on mathematical statistics and probability, Volume 1: Contributions to the theory of statistics, The Regents of the University of California.

Wold, H. (1966). Estimation of principal components and related models by iterative least squares. *Multivariate Analysis, 1*, 391–420.

Wold, H. (1973). *Nonlinear iterative partial least squares (NIPLAS) modelling: Some current developments*. New York: Academic.

Wold, H. (1974). Causal flows with latent variables: Partings of the ways in the light of NIPALS modelling. *European Economic Review, 5*(1), 67–86.

Wold, H. (1982). Systems under indirect observation using PLS. In C. Fornell (Ed.), *A second generation of multivariate analysis* (Vol. 1, pp. 325–347). New York, NY: Praeger.

Wright, S. (1921). Correlation and causation. *Journal of Agricultural Research, 20*(7), 557–585.

Wright, S. (1934). The method of path coefficients. *The Annals of Mathematical Statistics, 5*(3), 161–215.

Wright, S. (1960). Path coefficients and path regressions: alternative or complementary concepts? *Biometrics, 16*(2), 189–202.

Chapter 4
LISREL and Its Progeny

Sewall Wright's path coefficients were conceived as dimensionless binary indicators of whether a genetic trait was passed to an offspring, or not. Correlations in Wright's context were almost overkill, though their magnitude might have been considered to suggest varying degrees of confidence in heritability of a trait. As path analysis began to find application in analyzing relationships that were multivalued or continuous, limitations in the ability to resolve effects began to reveal themselves. It was in this context that Tukey (1954) advocated systematic replacement in path analysis of the dimensionless path coefficients by the corresponding concrete path regressions. Geneticists Turner and Stevens (1959) published their seminal paper presenting the mathematics of *path regression* and inviting extensions that would apply other developments in statistics. In the early days of data processing, both Herman Wold and his student Karl Jöreskog developed software, building on Turner and Stevens' mathematics, that took advantage of computer-intensive techniques becoming available to universities in the 1970s and 1980s.

4.1 LISREL

Wold's student, Karl Jöreskog, extended Wold and Lohmöller's (Lohmöller, 1988, 1989) methods in software implementations of covariance structure path analysis methods. Jöreskog's LISREL (an acronym for *LI*near *S*tructural *REL*ations) software was the early trend setter in computer-intensive path model, having appeared in a mainframe form in the late 1970s. Later in the 1980s (prior to the Windows 3.1 graphical interface) Lohmöller (1988) released desktop computing software that implemented both Jöreskog's as well as his own interpretation of Wold's ideas for path modeling in LVPLS. LISREL followed shortly with a desktop version, and these two packages set the standards for future path modeling software.

© Springer International Publishing Switzerland 2015
J.C. Westland, *Structural Equation Models*, Studies in Systems,
Decision and Control 22, DOI 10.1007/978-3-319-16507-3_4

Lohmöller's proponents have projected an ongoing animosity towards the LISREL and AMOS approaches. PLS path analysis is to this day considered to be a competitor of Jöreskog's LISREL approach based on maximum likelihood and the assumption of multivariate normality. For example Lauro and Vinzi (2002) complained:

> The goal of LISREL (or hard modeling) is actually to provide a statement of causality by seeking to find structurally or functionally invariant parameters, i.e. invariant features of the mechanism generating observable variables) that define how the world of interest to the model at hand works. These parameters are supposed to relate to causes describing the necessary relationships between variables within a closed system. Unfortunately, most often real data do not meet the requirements for this ideal.

Whether or not you accept Lauro and Vinzi's complaints about LISREL (and covariance structure modeling approaches in general) many researchers tend to be frustrated with the LISREL software for two main reasons:

1. It requires larger sample sizes, and simply does not compute path coefficients at all if the sample size is too small (we saw previously that PLS path analysis software addresses this through the dubious methods of resampling).
2. It requires that data be normally distributed, which is a problem with survey data that typically involves a distinctly non-normal 5- or 7-point Likert scale censored at zero.

Debates aside, Jöreskog initially focused on the corresponding concept for undirected graphs—a forest, or an undirected graph without cycles. Choosing an orientation for a forest produces a special kind of directed acyclic graph called a random tree, for which stochastic process modeling was reasonably well understood. Jöreskog based his confirmatory factor analysis (CFA) methods on random forests, ultimately extending these methods to the full linear structural relations (LISREL) software package introduced in the 1970s. LISREL was originally syntax based and mainframe based; it was promoted in marketing research by Claes Fornell at the University of Michigan.

Jöreskog's innovation was to conceive of path coefficients as covariances; thus one more solution concept to Wold's path models with latent variables comprised the first components of indicators (i.e., observations). Lohmöller's path analysis software included two methods for computing path coefficients between the latent variables (i.e., first principal components)—ordinary least squares regression—and covariance analysis.

Subsequent covariance method software has proven inconsistent in its methods for calculating path coefficients. It has even equivocated on whether path coefficients should be reported as covariances, as would be the product of covariance analysis; or whether these should be presented as correlations, consistent with Wright's original approach.

AMOS, a popular software package, now owned by IBM-SPSS, was developed by McArdle (J. J. McArdle, 1988; B. H. McArdle & Anderson, 2001; J. J. McArdle & Epstein, 1987; J. J. McArdle & Hamagami, 1996, 2001; J. J. McArdle &

McDonald, 1984), reformulates Jöreskog's mathematics in the more compact RAM format, and reports correlations, thus blurring the bounds between path analysis methodologies. Ultimately, though, as Wright (1960) observed, any particular format for reporting path coefficients can be recovered from the others, and the only important difference is in ease and utility of interpretation.

In all cases, though, covariance methods are a full information estimation method (all of the latent variables are simultaneously used in the path coefficient calculations) as opposed to Wright's or Wold's path analysis, which were limited estimation methods that computed the coefficient between each pair of latent variables individually and sequentially.

The basic method is to use the indicator values to estimate the covariance matrix. This matrix gives a unique entry for each path:

$$\begin{bmatrix} L_1 \to L_1 & \cdots & L_1 \to L_m \\ \vdots & \ddots & \vdots \\ L_m \to L_1 & \cdots & L_m \to L_m \end{bmatrix}$$

The diagonal contains the latent variable's variance, off-diagonal elements, and covariance between two latent variables. Some covariance software report correlations (a la classic Wright path coefficient) which they compute by dividing the path correlation by the square roots of the two diagonal terms (one for the row number, and the other for the column number).

Here is an example of the magnitude of the problem in an SEM context. Assume that your model has six (6) latent constructs (bubbles) (Fig. 4.1).

But you do not know a priori the precise relationships between constructs. In fact, with six constructs, what you can explore up front is a model with 15, not 5 links. And the number of links you have to explore goes up by the square of the number of bubbles (Fig. 4.2).

For discriminant analysis, you would assume that each one of these links can take a value of [+, 0, −] : Roughly, either the causal relationship (arrow) points forward, backward, or does not exist.

If you take Fig. 4.1 and label the latent variables 1 through 6, then the model implies that the only five entries in the covariance matrix are nonzero: $L_1 \to L_2$,

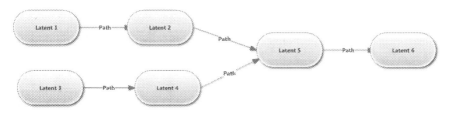

Fig. 4.1 Six constructs and five causal links

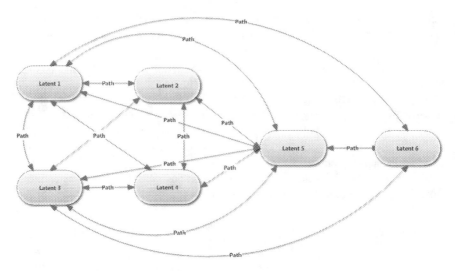

Fig. 4.2 Six constructs and $\dfrac{6(6-1)}{2} = 15$ potential causal relationships

$L_3 \rightarrow L_4$, $L_2 \rightarrow L_5$, $L_4 \rightarrow L_5$, and $L_5 \rightarrow L_6$. The other ten entries in the covariance matrix are restricted to be zero valued.

Jöreskog (1967) developed a rapidly converging iterative method for *exploratory* ML factor analysis (i.e., the factors are not defined in advance, but are discovered by exploring the solution space for the factors that explain the most variance) based on the Davidon-Fletcher-Powell math programming procedure commonly used in the solution of unconstrained nonlinear programs. As computing power evolved, other algorithms became feasible for searching the SEM solution space, and current software tends to use Gauss-Newton methods to optimize Browne's (Browne & Cudeck, 1989, 1992, 1993; Browne, Cudeck, Tateneni, & Mels, 2002) discrepancy function with an appropriate weight matrix that converges to ML, ULS, or GLS solutions for the SEM or to Browne's asymptotically distribution-free discrepancy function using polychoric (i.e., latent variables) correlations (Browne & Cudeck, 1989, 1992, 1993; Browne et al., 2002). Jöreskog (1969) extended this method to allow a priori specification of factors and factor loadings (i.e., the covariance of an unobserved factor and some observed "indicator") calling this *confirmatory* factor analysis. Overall fit of the a priori theorized model to the observed data could be measured by likelihood ratio techniques.

Jöreskog developed an early version of LISREL for *confirmatory factor analysis* (where latent factor relationships are correlations rather than causal; i.e., they do not have arrows). Later the method was extended to allow causality. In psychometrics and cognate fields, "structural equation modeling" (path modeling with latent variables) is sometimes used for causal inference and sometimes to get parsimonious descriptions of covariance matrices. For causal inference, questions of stability are central. If no causal inferences are made, stability under intervention is hardly relevant, nor are underlying equations "structural" in the econometric sense. The statistical assumptions (independence, distributions of error terms constant across subjects, parametric models for error distributions) would remain on the table. The confirmatory model testing provided by LISREL, AMOS, and other programs are the primary tools of descriptive analysis for hypothesis testing and theory confirmation for complex models with latent constructs.

Turner and Stevens' (1959) seminal paper introduced some of the more involved concepts of inner and outer models in path analysis structural equation modeling. Jöreskog's LISREL notation for structural equation models introduces a plethora of Greek symbols. The structural (latent variable) and measurement (indicator or measured factor) submodels are written in LISREL notation as (1) $\eta_i = B\eta_i + \Gamma\xi_i + \zeta_i$; (2) $y_i = \Lambda_y \eta_i + \zeta \epsilon_i$; and (3) $x_i = \Lambda_x \xi_i + \delta_i$. Furthermore in order to identify a LISREL model, parameters $B, \Gamma, \Lambda^x, \Lambda^y \Phi, \Psi, \Theta^\epsilon, \Theta^\delta$ (and more) have to be constrained by setting their values to 0, 1 or by setting various parameters to be equal. All of this is mind-boggling and Greek to most users.

Whether this notation reflects *physics envy*—the prejudice that anything researchable can be expressed in notation worthy of Newtonian mechanics—or merely excessive math enthusiasm, Jöreskog's Greek and index pushing does little to improve usefulness while leaving researchers mucking through a swamp of notation. A more civilized view is embraced by the authors of rival software AMOS (McArdle & McDonald, 1984, Arbuckle, Wothke, & SPSS Inc., 1999) who dispense with anything but path model diagrams in their AMOS-graphics software user interface, without losing expressability or generalizability in the ensuing statistical analysis (Blunch, 2008). AMOS (now an IBM product) advertises that they have dispensed with the equational minutia, and make it possible for researchers to concentrate on the path model.

There is a dedicated group of software packages which solely derived path coefficients through covariance structure modeling on normally distributed data. These packages have the advantage that they can generate goodness-of-fit statistics for the path model as a whole, though at the expense of only being able to process normal observations. With large enough datasets, it is argued that central limit theorem convergence will allow these methods to be used for well-behaved non-normal data. Unfortunately, as the LISREL and AMOS manuals observe, the increase in sample size may be several orders of magnitude.

Software packages for covariance structure modeling have been reviewed in several survey papers (Byrne, 2001; Dhrymes, 1974; Hox, 1995; Lydtin, Lohmöller, Lohmöller, Schmitz, & Walter, 1975; Marsh, Byrne, & Craven, 1992). Hox (1995) asserts that the significant contrasts appear in fit statistics, both number and applicability to specific problems. John Fox (2006a, 2006b) argues that many of these fit statistics are ad hoc and difficult to interpret.

Table 4.1 Covariance structure modeling software

Software	Methods supported
AMOS (IBM)	Structural equation models, multiple regression, multivariate regression, confirmatory factor analysis, structured means analysis, path analysis, and multiple population comparisons. Many consider the GUI to be the best of the commercial covariance method packages. Developed by James Arbuckle, now part of IBM through its purchase of SPSS
CALIS (SAS)	A SAS Proc which implements multiple and multivariate linear regression; linear measurement-error models; path analysis and causal modeling; simultaneous equation models with reciprocal causation; exploratory and confirmatory factor analysis of any order; canonical correlation; a wide variety of other (non)linear latent variable models CALIS (Hartmann, 1992)
EQS (MSI)	Structural equation models, multiple regression, multivariate regression, confirmatory factor analysis, structured means analysis, path analysis, and multiple population comparisons (Bentler, 1985, 1995)
LISREL (SSI)	Evolved from Karl Jöreskog's branch of algorithm development as a student of Herman Wold, LISREL (Jöreskog & Sörbom, 1989, 1996) was the first computer-based covariance structure modeling package (implemented on mainframes in the late 1970s)
Mplus (Muthén & Muthén)	Exploratory factor analysis; structural equation modeling; item response theory analysis; growth modeling; mixture modeling (latent class analysis); longitudinal mixture modeling (hidden Markov, latent transition analysis, latent class growth analysis, growth mixture analysis); survival analysis (continuous- and discrete-time); multilevel analysis; complex survey data analysis; Bayesian analysis; Monte Carlo simulation
OpenMx/OpenSEM (Virginia)	Cross platform Mac OS X, Windows XP, Windows Vista, and several varieties of Linux; open source with integration with R statistical language; covariance modeling with means; missing data; categorical threshold estimation; hierarchical model definition; matrix algebra calculations; user-specified functions for model specification; user-specified objective functions; community Wiki and forums
sem (R)	This is an R-code procedure that implements the RAM formulation of covariance structure models
TETRAD (CMU)	Cross platform Mac OS X, Windows XP, Windows Vista, and several varieties of Linux; open source, community forums

In another option the *sem* procedure in R-language uses the reticular action model (RAM) formulation (Fox, 2002, 2006a; Joreskog & Van Thillo, 1972) of covariance structure models. To illustrate, *sem*'s author John Fox reformulates the classic SEM model of Blau, Duncan, and Tyree (1967). Fox also implements, in his SEM software, the reticular action model (RAM) formulation of B. H. McArdle and Anderson (2001), and Sobel (1982), dispensing which the plethora of Greek notation that he feels overly complicates Jöreskog's formulation (Table 4.1).

4.2   LISREL Performance Statistics

A cottage industry in ad hoc fit indices and their evaluation has developed around covariance structure methods. It should be noted up front that a "good fit" is not the same as strength of relationship: one could have perfect fit when all variables in the model were totally uncorrelated, as long as the researcher does not instruct the SEM software to constrain the variances. In fact, the lower the correlations stipulated in the model, the easier it is to find "good fit." The stronger the correlations, the more power SEM has to detect an incorrect model. When correlations are low, the researcher may lack the power to reject the model at hand. Also, all measures over-estimate goodness of fit for small samples, though RMSEA and CFI are less sensitive to sample size than others (Kleinberg, 2000).

In cases where the variables have low correlation, the structural (path) coefficients will also be low. Researchers should report not only goodness-of-fit measures but also the structural coefficients so that the strength of paths in the model can be assessed. Likewise, one can have good fit in a misspecified model. One indicator of this occurring is if there are high modification indexes (MI) in spite of good fit. High MIs indicate multicollinearity in the model and/or correlated error.

All other things equal, a model with fewer indicators per factor will have a higher apparent fit than a model with more indicators per factor. Fit coefficients that reward parsimony are one way to adjust for this tendency (Table 4.2).

Table 4.2 Performance statistics

Performance statistic	Description, use, and advantages/disadvantages
Root mean square residuals, or RMS residuals, or RMSR, or RMR	The closer the RMR to 0 for a model being tested, the better the model fit. RMR is the coefficient which results from taking the square root of the mean of the squared residuals, which are the amounts by which the sample variances and covariances differ from the corresponding estimated variances and covariances, estimated on the assumption that your model is correct. Fitted residuals result from subtracting the sample covariance matrix from the fitted or estimated covariance matrix. LISREL computes RMSR. AMOS does also, but calls it RMR.
Standardized root mean square residual, standardized RMR (SRMR)	The smaller the standardized RMR, the better the model fit. SRMR is the average difference between the predicted and observed variances and covariances in the model, based on standardized residuals. Standardized residuals are fitted residuals (see above) divided by the standard error of the residual (this assumes a large enough sample to assume stability of the standard error). SRMR is 0 when model fit is perfect.

(continued)

Table 4.2 (continued)

Performance statistic	Description, use, and advantages/disadvantages
Model chi-square	Model chi-square, also called discrepancy or the discrepancy function, is the most common fit test, printed by all computer programs. AMOS outputs it as CMIN. The chi-square value should not be significant if there is a good model fit, while a significant chi-square indicates lack of satisfactory model fit. That is, chi-square is a "badness of fit" measure in that a finding of significance means that the given model's covariance structure is significantly different from the observed covariance matrix. If model chi-square <0.05, the researcher's model is rejected. LISREL refers to model chi-square simply as chi-square, but synonyms include the chi-square fit index, chi-square goodness of fit, and chi-square badness of fit. Model chi-square approximates for large samples what in small samples and log-linear analysis is called G^2, the generalized likelihood ratio.
	There are three ways, listed below, in which the chi-square test may be misleading. Because of these reasons, many researchers who use SEM believe that with a reasonable sample size (e.g., >200) and good approximate fit as indicated by other fit tests (e.g., NNFI, CFI, RMSEA, and others discussed below), the significance of the chi-square test may be discounted and that a significant chi-square is not a reason by itself to modify the model.
	The more complex the model, the more likely a good fit. In a just-identified model, with as many parameters as possible and still achieving a solution, there will be a perfect fit. Put another way, chi-square tests the difference between the researcher's model and a just-identified version of it, so the closer the researcher's model is to being just-identified, the more likely the good fit will be found.
	The larger the sample size, the more likely the rejection of the model and the more likely a type II error (rejecting something true). In very large samples, even tiny differences between the observed model and the perfect-fit model may be found significant.
	The chi-square fit index is also very sensitive to violations of the assumption of multivariate normality. When this assumption is known to be violated, the researcher may prefer *Satorra-Bentler scaled chi-square*, which adjusts model chi-square for non-normality.
Hoelter's critical N	Is the size the sample size must reach for the researcher to accept the model by chi-square, at the 0.05 or 0.01 levels. This throws light on the chi-square fit index's sample size problem. Hoelter's N should be greater than 200.
Satorra-Bentler scaled chi-square	Sometimes called Bentler-Satorra chi-square, this is an adjustment to chi-square which penalizes chi-square for the amount of kurtosis in the data. That is, it is an adjusted chi-square statistic which attempts to correct for the bias introduced when data are markedly non-normal in distribution.

(continued)

Table 4.2 (continued)

Performance statistic	Description, use, and advantages/disadvantages
Relative chi-square, also called normal chi-square	Is the chi-square fit index divided by degrees of freedom, in an attempt to make it less dependent on sample size. Carmines and McIver (1981) state that relative chi-square should be in the 2:1 or 3:1 range for an acceptable model. Some researchers allow values as high as 5 to consider a model adequate fit, while others insist relative chi-square be 2 or less. AMOS lists relative chi-square as CMIN/DF.
Goodness-of-fit index, GFI (Jöreskog-Sörbom GFI)	GFI varies from 0 to 1, but theoretically can yield meaningless negative values. A large sample size pushes GFI up. Though analogies are made to R-square, GFI cannot be interpreted as percent of error explained by the model. Rather it is the percent of observed covariances explained by the covariances implied by the model. That is, R2 in multiple regression deals with error variance whereas GFI deals with error in reproducing the variance-covariance matrix. As GFI often runs high compared to other fit models, some suggest using 0.95 as the cutoff. By convention, GFI should by equal to or greater than 0.90 to accept the model. LISREL and AMOS both compute GFI.
Adjusted goodness-of-fit index, AGFI	AGFI is a variant of GFI which adjusts GFI for degrees of freedom: the quantity $(1 - GFI)$ is multiplied by the ratio of your model's df divided by df for the baseline model, and then AGFI is 1 minus this result. AGFI can yield meaningless negative values. AGFI > 1.0 is associated with just-identified models and models with almost perfect fit. AGFI < 0 is associated with models with extremely poor fit, or based on small sample size. AGFI should also be at least 0.90. Like GFI, AGFI is also biased downward when degrees of freedom are large relative to sample size, except when the number of parameters is very large. Like GFI, AGFI tends to be larger as sample size increases; correspondingly, AGFI may underestimate fit for small sample sizes, according to Bollen (1990).
	The goodness-of-fit index (GFI) and the adjusted goodness-of-fit index (AGFI) are ad hoc measures of the descriptive adequacy of the model. Although the GFI and AGFI are thought of as proportions, comparing the value of the fitting criterion for the model with the value of the fitting criterion when no model is fit to the data, these indices are not constrained to the interval 0–1. Several rough cutoffs for the GFI and AGFI have been proposed; a general theme is that they should be close to 1. It is probably fair to say that the GFI and AGFI are of little practical value.
Centrality index, CI	CI is a function of model chi-square, degrees of freedom in the model, and sample size. By convention, CI should be 0.90 or higher to accept the model.
Noncentrality parameter, NCP	This is also called the McDonald noncentrality parameter index and DK, is chi-square penalizing for model complexity. To force it to scale to 1, the conversion is exp(−DK/2). NCP is used with a table of the noncentral chi-square distribution to assess power. RMSEA, CFI, RNI, and CI are related to the noncentrality parameter. Raykov (2005) has argued that fit measures based on noncentrality are biased.

(continued)

Table 4.2 (continued)

Performance statistic	Description, use, and advantages/disadvantages
Goodness-of-fit tests comparing the given model with an alternative model	This set of goodness-of-fit measures compare your model to the fit of another model. This is well and good if there is a second model. When none is specified, statistical packages usually default to comparing your model with the independence model, or even allow this as the only option. The independence model is the null model, which is the model in which variables are assumed to be uncorrelated with the dependent(s). Since the fit of the independence model is usually terrible, comparing your model to it will generally make your model look good but may not serve your research purposes.
The comparative fit index, CFI	CFI is also known as the Bentler comparative fit index. CFI compares the existing model fit with a null model which assumes that the latent variables in the model are uncorrelated (the "independence model"). That is, it compares the covariance matrix predicted by the model to the observed covariance matrix, and compares the null model (covariance matrix of 0s) with the observed covariance matrix, to gauge the percent of lack of fit which is accounted for by going from the null model to the researcher's SEM model. Note that to the extent that the observed covariance matrix has entries approaching 0s, there will be no nonzero correlation to explain and CFI loses its relevance. CFI is similar in meaning to NFI (see below) but penalizes for sample size. CFI and RMSEA are among the measures least affected by sample size (Fan et al., 1999). CFI varies from 0 to 1. CFI close to 1 indicates a very good fit. CFI is also used in testing modifier variables (those which create a heteroscedastic relation between an independent and a dependent, such that the relationship varies by class of the modifier). By convention, CFI should be equal to or greater than 0.90 to accept the model, indicating that 90 % of the covariation in the data can be reproduced by the given model.
GFI based on predicted vs. observed covariances, penalizing lack of parsimony	Parsimony measures: These measures penalize for lack of parsimony, since more complex models will, all other things equal, generate better fit than less complex ones. They do not use the same cutoffs as their counterparts (e.g., PCFI does not use the same cutoff as CFI) but rather will be noticeably lower in most cases. Used when comparing models, the higher parsimony measure represents the better fit.
Parsimony ratio (PRATIO)	PRATO is the ratio of the degrees of freedom in your model to degrees of freedom in the independence (null) model. PRATIO is not a goodness-of-fit test itself, but is used in goodness-of-fit measures like PNFI and PCFI which reward parsimonious models (models with relatively few parameters to estimate in relation to the number of variables and relationships in the model). See also the parsimony index, below.
Parsimony index	The parsimony index is the parsimony ratio times BBI, the Bentler/Bonnett index, discussed above. It should be greater than 0.9 to assume good fit.

(continued)

Table 4.2 (continued)

Performance statistic	Description, use, and advantages/disadvantages
Root mean square error of approximation	RMSEA is also called RMS or RMSE or discrepancy per degree of freedom. By convention, there is good model fit if RMSEA is less than or equal to 0.05. There is adequate fit if RMSEA is less than or equal to 0.08. More recently, Hu and Bentler (1999) have suggested RMSEA ≤ 0.06 as the cutoff for a good model fit. RMSEA is a popular measure of fit, partly because it does not require comparison with a null model and thus does not require the author posit as plausible a model in which there is complete independence of the latent variables as does, for instance, CFI. Also, RMSEA has a known distribution, related to the noncentral chi-square distribution, and thus does not require bootstrapping to establish confidence intervals. Confidence intervals for RMSEA are reported by some statistical packages. It is one of the fit indexes less affected by sample size, though for smallest sample sizes it overestimates goodness of fit (Fan et al., 1999).
Goodness-of-fit measures based on information theory	Measures in this set are appropriate when comparing models which have been estimated using maximum likelihood estimation. As a group, this set of measures is less common in the literature, but that is changing. All are computed by AMOS. They do not have cutoffs like 0.90 or 0.95. Rather they are used in comparing models, with the lower value representing the better fit.
Akaike Information Criterion (AIC)	AIC is a goodness-of-fit measure which adjusts model chi-square to penalize for model complexity (that is, for overparameterization). Thus AIC reflects the discrepancy between model-implied and observed covariance matrices. Unlike model chi-square, AIC may be used to compare nonhierarchical as well as hierarchical (nested) models, whereas model chi-square difference is used only for the latter. It is possible to obtain AIC values <0. AIC close to 0 reflects good fit and between two AIC measures, the lower one reflects the model with the better fit. AIC can also be used for hierarchical (nested) models, as when one is comparing nested modifications of a model. In this case, one stops modifying when AIC starts rising. Burnham and Anderson (2002) provide further information on AIC and related information theory measures.
Consistent AIC (CAIC)	Consistent AIC penalizes for sample size as well as model complexity (lack of parsimony). The penalty is greater than AIC or BCC but less than BIC. As with AIC, the lower the CAIC measure, the better the fit.
Browne-Cudeck criterion (BCC)	The Browne-Cudeck criterion is also called the Cudeck and Browne single-sample cross-validation index. It should be close to 0.9 to consider fit good.
Expected cross-validation index	ECVI in its usual variant is equivalent to BCC, and is useful for comparing non-nested models. Like AIC, it reflects the discrepancy between model-implied and observed covariance matrices. Lower ECVI is better fit. When comparing nested models, chi-square difference is normally used. ECVI used for nested models differs from chi-square difference in that ECVI penalizes for number of free parameters. This difference between ECVI and chi-square difference could affect conclusions if the chi-square difference is a substantial relative to degrees of freedom. MECVI is a variant on BCC, differing in scale factor.

(continued)

Table 4.2 (continued)

Performance statistic	Description, use, and advantages/disadvantages
Bayesian Information Criterion (BIC)	BIC is the Bayesian Information Criterion, also known as Akaike's Bayesian Information Criterion (ABIC) and the Schwarz Bayesian Criterion (SBC). Like CAIC, BIC penalizes for sample size as well as model complexity. In general, BIC has a conservative bias tending towards type II error (thinking there is poor model fit when the relationship is real). Put another way, compared to AIC, BCC, or CAIC, BIC more strongly favors parsimonious models with fewer parameters. BIC is recommended when sample size is large or the number of parameters in the model is small.
	BIC actually has a sound statistical basis (Raftery & Hout, 1993). The BIC adjusts the likelihood-ratio chi-square statistic L2 for the number of parameters in the model, number of observed variables, and sample size. Negative values of BIC indicate a model that has greater support from the data than the just-identified model, for which BIC is 0. Differences in BIC may be used to compare alternative overidentified models; indeed, the BIC is used in a variety of contexts for model selection, not just in structural equation modeling. Raftery and Hout (1993) suggest that a BIC difference of 5 is indicative of "strong evidence" that one model is superior to another, while a difference of 10 is indicative of "conclusive evidence."
	BIC is an approximation to the log of a Bayes factor for the model of interest compared to the saturated model. BIC became popular in sociology after it was popularized by Raftery in the 1980s (Raftery & Hout, 1993). Winship and Morgan (1999) identify caveats for models based on a large sample size but which have little variance in their variables and/or highly collinear independents; these may yield misleading model fit using BIC.

References

Arbuckle, J. L., Wothke, W., & SPSS Inc. (1999). *Amos 4.0 users' guide*. Chicago, IL: SmallWaters Corporation, SPSS Inc.

Bentler, P. M. (1985). *Theory and implementation of EQS: A structural equations program*. Los Angeles, CA: BMDP Statistical Software.

Bentler, P. M. (1995). *EQS structural equations program manual*. Encino, CA: Multivariate Software.

Blau, P. M., Duncan, O. D., & Tyree, A. (1967). The process of stratification. In D. B. Grusky (Ed.), *Social stratification. Class, race & gender, in sociological perspective* (pp. 317–329). San Francisco, CA: Boulder.

Blunch, N. J. (2008). *Introduction to structural equation modelling using SPSS and AMOS*. London, UK: Sage Publications.

Bollen, K. A. (1990). Overall fit in covariance structure models: Two types of sample size effects. *Psychological Bulletin, 107*(2), 256.

Browne, M. W., & Cudeck, R. (1989). Single sample cross-validation indices for covariance structures. *Multivariate Behavioral Research, 24*(4), 445–455.

Browne, M. W., & Cudeck, R. (1992). Alternative ways of assessing model fit. *Sociological Methods & Research, 21*(2), 230–258.

Browne, M. W., & Cudeck, R. (1993). Alternative ways of assessing model fit. *Sage Focus Editions, 154*, 136.

Browne, M. W., Cudeck, R., Tateneni, K., & Mels, G. (2002). CEFA: Comprehensive exploratory factor analysis. *Computer.*

Browne, M. W., Cudeck, R., Tateneni, K., & Mels, G. (2004). *CEFA: Comprehensive exploratory factor analysis, Version 2.00 [Computer software and manual].* Columbus: The Ohio State University.

Burnham, K. P., & Anderson, D. R. (2002). *Model selection and multimodel inference: A practical information-theoretic approach.* New York, NY: Springer Science & Business Media.

Byrne, B. M. (2001). Structural equation modeling with AMOS, EQS, and LISREL: Comparative approaches to testing for the factorial validity of a measuring instrument. *International Journal of Testing, 1*(1), 55–86.

Carmines, E. G., & McIver, J. P. (1981). Analyzing models with unobserved variables: Analysis of covariance structures. In G. Bohrnstedt & E. Borgatta (Eds.), *Social measurement: Current issues* (pp. 65–115). Beverly Hills, CA: Sage.

Dhrymes, P. J. (1974). *Econometrics.* New York, NY: Springer.

Fan, X., Thompson, B., & Wang, L. (1999). Effects of sample size, estimation methods, and model specification on structural equation modeling fit indexes. *Structural Equation Modeling, 6*(1), 56–83.

Fox, J. (2002). Structural equation models. *CRAN website.*

Fox, J. (2006a). Structural equation modeling with the sem package in R. *Structural Equation Modeling, 13*(3), 465–486.

Fox, J. (2006b). Teacher's corner: Structural equation modeling with the sem package in R. *Structural Equation Modeling, 13*(3), 465–486.

Fox, J., Nie, Z., & Byrnes, J. (2012). Sem: Structural equation models. R package version 3.0. (sem 3.0).

Hartmann, W. M. (1992). *The CALIS procedure: Extended user's guide.* Cary, NC: SAS Institute.

Hox, J. J. (1995). Amos, EQS, and LISREL for Windows: A comparative review. *Structural Equation Modeling, 2*(1), 79–91.

Hu, L.-t., & Bentler, P. M. (1999). Cutoff criteria for fit indexes in covariance structure analysis: Conventional criteria versus new alternatives. *Structural Equation Modeling, 6*(1), 1–55.

Jöreskog, K. G. (1967). A general approach to confirmatory maximum likelihood factor analysis. *ETS Research Bulletin Series, 1967*(2), 183–202.

Jöreskog, K. G. (1969). A general approach to confirmatory factor analysis. *Psychometrika, 34*, 183–202.

Jöreskog, K. G., & Sörbom, D. (1989). *LISREL 7: A guide to the program and applications.* Chicago, IL: SPSS.

Jöreskog, K. G., & Sörbom, D. (1996). *LISREL 8 user's reference guide.* Chicago, IL: Scientific Software International.

Joreskog, K. G., & Van Thillo, M. (1972). LISREL: A general computer program for estimating a linear structural equation system involving multiple indicators of unmeasured variables. *ETS Research Bulletin Series, 1972*(2), 1–71.

Kleinberg, J. M. (2000). Navigation in a small world. *Nature, 406*(6798), 845.

Lauro, C., & Vinzi, V. E. (2002). Some contributions to PLS Path Modeling and a system for the European Customer Satisfaction. *Universita di Milano Bicocca, Milano, atti della XLI riunione scientifica SIS.*

Lohmöller, J.-B. (1988). The PLS program system: Latent variables path analysis with partial least squares estimation. *Multivariate Behavioral Research, 23*(1), 125–127.

Lohmöller, J.-B. (1989). *Latent variable path modeling with partial least squares.* Heidelberg, Germany: Physica.

Lydtin, H., Lohmöller, G., Lohmöller, R., Schmitz, H., & Walter, I. (1975). *Hemodynamic studies on Adalat in healthy volunteers and in patients.* Paper presented at the 2nd International Adalat® Symposium.

Marsh, H. W., Byrne, B. M., & Craven, R. (1992). Overcoming problems in confirmatory factor analyses of MTMM data: The correlated uniqueness model and factorial invariance. *Multivariate Behavioral Research, 27*(4), 489–507.

McArdle, J. J. (1988). Dynamic but structural equation modeling of repeated measures data. In J. R. Nesselroade & R. B. Cattell (Eds.), *Handbook of multivariate experimental psychology* (pp. 561–614). New York, NY: Springer.

McArdle, B. H., & Anderson, M. J. (2001). Fitting multivariate models to community data: A comment on distance-based redundancy analysis. *Ecology, 82*(1), 290–297.

McArdle, J. J., & Epstein, D. (1987). Latent growth curves within developmental structural equation models. *Child Development, 58*, 110–133.

McArdle, J. J., & Hamagami, F. (1996). Multilevel models from a multiple group structural equation perspective. In G. A. Marcoulides & R. E. Schumacker (Eds.), *Advanced structural equation modeling: Issues and techniques* (pp. 89–124). Mahwah, NJ: Erlbaum.

McArdle, J. J., & Hamagami, F. (2001). Latent difference score structural models for linear dynamic analyses with incomplete longitudinal data. In L. M. Collins & A. G. Sayer (Eds.), *New methods for the analysis of change. Decade of behavior* (pp. 139–175). Washington, DC: American Psychological Association.

McArdle, J. J., & McDonald, R. P. (1984). Some algebraic properties of the reticular action model for moment structures. *British Journal of Mathematical and Statistical Psychology, 37*(2), 234–251.

Raftery, A. E., & Hout, M. (1993). Maximally maintained inequality: Expansion, reform, and opportunity in Irish education, 1921–75. *Sociology of Education, 66*, 41–62.

Raykov, T. (2005). Analysis of longitudinal studies with missing data using covariance structure modeling with full-information maximum likelihood. *Structural Equation Modeling, 12*(3), 493–505.

Sobel, M. E. (1982). Asymptotic confidence intervals for indirect effects in structural equation models. *Sociological Methodology, 13*(1982), 290–312.

Tukey, J. W. (1954). Causation, regression, and path analysis. In O. Kempthorne, T. A. Bancroft, J. W. Gowen, & J. L. Luch (Eds.), *Statistics and mathematics in biology* (pp. 35–66). Ames, IA: Iowa State College Press.

Turner, M. E., & Stevens, C. D. (1959). The regression analysis of causal paths. *Biometrics, 15*(2), 236–258.

Winship, C., & Morgan, S. L. (1999). The estimation of causal effects from observational data. *Annual Review of Sociology, 25*, 659–706.

Wright, S. (1960). Path coefficients and path regressions: Alternative or complementary concepts? *Biometrics, 16*(2), 189–202.

Chapter 5
Systems of Regression Equations

5.1 The Birth of Structural Equation Modeling

Alfred Cowles III hailed from an established Chicago publishing family, his father and uncle having founded the *Chicago Tribune* and *Cleveland Leader*, respectively (Grier, 2013). For a short time after WWI Cowles successfully ran a Chicago investment firm that acquired and restructured small railroads. His firm also published a stock market newsletter providing fundamental analysis and recommendations on railroad stock issues as well as other investments, and for a time there was even an Alfred Cowles Railroad.

Diagnosed with tuberculosis in the late 1920s, Cowles consolidated his investments (just prior to the 1929 crash) and moved to Colorado Springs in search of better health (Grier, 2013). Consigned to a life of enforced leisure, he filled his time developing linear regression models that simultaneously compared the predictions of 24 stock market newsletters to actual stock performance. Cowles quickly came to the conclusion that forecasters were guessing, they offered little useful investment information, and were more often wrong than right (Cowles, 1933). Understandably, he also applied his regression skills to investigate whether good climates, like Colorado Springs, improved the outcome of tuberculosis (Cowles & Chapman, 1935) with somewhat more optimistic results.

The pen and paper calculation required at the time for the regression formulas he used soon exceeded his capabilities as a lone researcher. At this point he made a decision to invest some of his fortune to create the Cowles Commission, an institution dedicated to linking economic theory to mathematics and statistics. To that end, its mission was to develop a specific, probabilistic framework for estimating simultaneous regression equations to model the US economy.

The Cowles Commission moved from Colorado Springs to the University of Chicago in 1939 where economist Tjalling Koopmans (1951, 1957) developed the systems of regression tools that Cowles originally had sought. This period also expanded Cowles personal files into what ultimately became the Compustat and CRSP databases, and created the market index that eventually became the Standard and Poor's 500 Index. Throughout its 15 years at the University of Chicago, the Commission clashed repeatedly with the Economics Department and in 1955 ultimately made the decision to move to Cowles' *alma mater* Yale, where it was renamed the Cowles Foundation.

The Cowles Commission's most important contribution to statistics was in exposing the bias of ordinary least squares regression coefficient estimates. Cowles researchers developed new methods such as the indirect least squares, instrumental variable methods, full information maximum likelihood method, and limited information maximum likelihood methods to resolve this problem (Christ, 1994).

Eleven Cowles associates ultimately received the Nobel Prize in Economics, most notably (for this book) Trygve Haavelmo, who introduced his Scandinavian colleagues Herman Wold and Karl Jöreskog to Cowles' simultaneous regression equation approaches. Wold ultimately went on to develop PLS-PA and Jöreskog developed LISREL as latent variable alternatives that they considered more suitable for the abstract and unstructured problems of sociology, education, and psychology.

5.2   Simultaneous Regression Equation Models

While Wold and Joreskog were pursuing idiosyncratic solutions to path coefficients, work at the Cowles Commission, under Koopmans, Zellner, Anderson, Dhrymes, and many others, made rapid progress in devising econometric tools for the networked relationships found in the US economy. Their simultaneous equation regression (also "systems of regression equation") approaches now comprise the mainstream approach in econometrics and other fields for mapping network relationships between variables. In general, this line of research has eschewed working with latent variables, but only because there was no need for special methods for dealing with them—they are linear functions of indicators, with coefficients set by the first principal component. We explore the use of systems of regression equation approaches with latent variables later in this chapter.

Simultaneous equation models are a multi-equation regression model in the form of a set of linear simultaneous equations, where the covariance matrix is not diagonal (i.e., there is covariance between the separate linear equations). It is extremely common in econometrics to encounter systems of regression equations (which need to be estimated simultaneously, i.e., as a network of relationships across equations). The equations are written in vector-matrix form, and all endogenous variables are algebraically moved to the left-hand side to produce what is called the "structural form" system of equations. It was this usage of structural form that was adopted by Wold and Jöreskog to describe their particular setups, which is why they called them structural equation models (SEM). Further algebraic manipulation to pull all endogenous variables to the left-hand side and exogenous variables to the right-hand side of the equation produces the "reduced form" system of equations. The reduced form is a simple general linear model which may be estimated using ordinary least squares regression.

Unfortunately, the task of decomposing the estimated matrix algebraically into the individual factors is often complicated. There may indeed be questions concerning whether the estimators for the original equation can be algebraically recovered and are unique—it may be possible to have no solutions, or to have an infinite number of solutions derived from the reduced form. To assure that the estimators we recover from the reduced form are unique, we apply specific identification conditions before estimating. If these are not met, a model restructuring is required.

In order for a unique estimate to be derived, three conditions must be met:

1. The error terms are assumed to be serially independent and identically distributed.
2. The rank of the matrix of exogenous regressors must be equal to the number of exogenous regressors.
3. The identification conditions require that the number of unknowns in this system of equations not exceed the number of equations. There are two identification conditions: (1) the *order condition* requires that the number of excluded exogenous variables is greater or equal to the number of included endogenous variables; and (2) the *rank condition* states put constraints on the rank of the matrix which is obtained from the reduced form exogenous coefficient matrix by crossing out those columns which correspond to the excluded endogenous variables, and those rows which correspond to the included exogenous variables.

Path analysis using linear simultaneous equations has the following advantages:

1. They describe path coefficients in terms of regression coefficients (Tukey (1954) claimed that they were more informative than correlations, and easier to interpret than covariances).
2. They are full information methods (versus PLS path analysis which is limited information).
3. They have well-defined performance metrics (fit statistics) and analysis approaches, including residual analysis for underlying model assumptions; neither PLS path analyses nor covariance methods have this. In particular, hypothesis tests are well defined, and can convincingly reject alternative hypotheses.
4. They allow for residuals that can be plotted and inspected for data problems such as autocorrelation, heteroskedacity, non-normality, outliers, and more. The two other approaches do not allow this—(1) PLS path analysis obscures any analysis of this sort because of resampling; and (2) iterative search algorithms that underlie covariance solutions obscure the impact of non-normal and problem data on residuals.
5. There are transformations that are well understood (logit, probit, log, Box-Cox, etc.) that can be used on nonlinear data.

Consider an example of a reformulation of a latent variable structural model in a fashion that allows systems of equation estimation of a path model (Fig. 5.1).

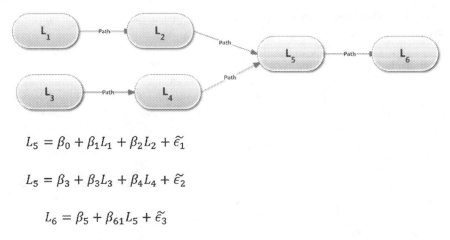

$$L_5 = \beta_0 + \beta_1 L_1 + \beta_2 L_2 + \widetilde{\epsilon_1}$$

$$L_5 = \beta_3 + \beta_3 L_3 + \beta_4 L_4 + \widetilde{\epsilon_2}$$

$$L_6 = \beta_5 + \beta_{61} L_5 + \widetilde{\epsilon_3}$$

Fig. 5.1 Latent variable path model and estimated system of equations

The reduced form of this system can be estimated with OLS regression, and the original parameters (and thus path coefficients on the structural model) can be recovered by reversing the algebraic transformation that yielded the reduced form. The standard identification conditions (rank and order) apply for identification (Greene & Zhang, 2003). When a model is identified, this means that unique estimates for the path coefficients can be obtained; over- or under-identification results in either multiple estimates for each path or none at all.

5.3   Estimation

The most common estimation methods for the simultaneous equation models are the following:

1. *Two-stage least squares* method, developed independently by Theil (1953) and R. L. Basmann (1957, 1963): It is an equation-by-equation technique, where the endogenous regressors on the right-hand side of each equation are being instrumented with the regressors from all other equations.
2. *Indirect least squares* is an approach in econometrics where the coefficients in a simultaneous equation model are estimated from the reduced form model using ordinary least squares. For this, the structural system of equations is transformed into the reduced form first. Once the coefficients are estimated the model is put back into the structural form.
3. The *"limited information" maximum likelihood* method was suggested by C. A. Anderson (1983), J. C. Anderson and Gerbing (1988), and T. W. Anderson and Rubin (1949, 1950).

Table 5.1 Systems of equation regression software

Software	Features
systemfit (R)	R's *systemfit* procedure can estimate systems of linear equations within the R programming environment, and can be used for "ordinary least squares" (OLS), "seemingly unrelated regression" (SUR), and the instrumental variable (IV) methods "two-stage least squares" (2SLS) and "three-stage least squares" (3SLS), where SUR and 3SLS estimations can optionally be iterated. Furthermore, the systemfit package provides tools for several statistical tests. It has been tested on a variety of datasets and its reliability is demonstrated
SAS (SAS)	PROC *MODEL* estimates: ARIMA, PDL, dynamic modeling; supports the following methods for parameter estimation: (1) ordinary least squares (OLS); (2) two-stage least squares (2SLS); (3) seemingly unrelated regression (SUR) and iterative SUR (ITSUR); (4) three-stage least squares (3SLS) and iterative 3SLS (IT3SLS); (5) generalized method of moments (GMM); (6) simulated method of moments (SMM); (7) full information maximum likelihood (FIML); (8) general log-likelihood maximization; (9) simulation and forecasting capabilities; (9) Monte Carlo simulation; and (10) goal-seeking solutions
STATA (Stata)	STATA's *reg3* Command estimates OLS, 2SLS, and 3SLS with some limitations
SPSS/Systat/AMOS (IBM)	Neither SPSS nor Systat packages support estimation of 3SLS or FIML; AMOS package estimates 2SLS
Eviews (HIS)	Windows-based econometric and forecasting software; has object-oriented interface to powerful statistical, forecasting, and modeling tools
LIMDEP (ESI)	Single-equation and simultaneous-equation regression models
MATLAB, Octave, Gauss, and Excel	Computational software that is sometimes redeployed for simultaneous equation regression analysis

4. The *three-stage least squares* estimator was introduced by Zellner (1962) and Zellner and Theil (1962). It combines two-stage least squares (2SLS) with seemingly unrelated regressions (SUR). There are variations on the method, including i-3SLS which involves an iterative search for estimators.

5. A *seemingly unrelated regression (SUR)* estimation procedure may be used if the error terms are not independent. Seemingly unrelated regressions consist of several regression equations, each having its own dependent variable and potentially different sets of exogenous explanatory variables. Equation-by-equation estimates are consistent, however generally not as efficient as the SUR method. When the covariance matrix is known to be diagonal, there are no cross-equation correlations between the error terms. In this case the system becomes not seemingly but truly unrelated. Tables 5.1 and 5.2 summarize the software available for systems of regression analysis and their performance metrics respectively.

Table 5.2 Performance metrics for systems methods

Performance metric	Application
R-squared	In statistics, the coefficient of determination R-squared is used in the context of statistical models whose main purpose is the prediction of future outcomes on the basis of other related information. It is the proportion of variability in a dataset that is accounted for by the statistical model. It provides a measure of how well future outcomes are likely to be predicted by the model. There are several different definitions of R-squared which are only sometimes equivalent. One class of such cases includes that of linear regression. In this case, if an intercept is included then R-squared is simply the square of the sample correlation coefficient between the outcomes and their predicted values, or in the case of simple linear regression, between the outcomes and the values of the single regressor being used for prediction. In such cases, the coefficient of determination ranges from 0 to 1. Important cases where the computational definition of R-squared can yield negative values, depending on the definition used, arise where the predictions which are being compared to the corresponding outcomes have not been derived from a model-fitting procedure using those data, and where linear regression is conducted without including an intercept. Additionally, negative values of R-squared may occur when fitting nonlinear trends to data. In these instances, the mean of the data provides a fit to the data that is superior to that of the trend under this goodness-of-fit analysis
F-test	An F-test is any statistical test in which the test statistic has an F-distribution under the null hypothesis. It is most often used when comparing statistical models that have been fit to a dataset, in order to identify the model that best fits the population from which the data were sampled. Exact F-tests mainly arise when the models have been fit to the data using least squares
t-Statistics (on individual parameters)	The t-statistic is a ratio of the departure of an estimated parameter from its notional value and its standard error. It is used in hypothesis testing, for example in the Student's t-test, in the augmented Dickey–Fuller test
Graphical examination of plots of regression residuals	Flexible ad hoc method of testing model assumptions: The following four assumptions on the random errors are equivalent to the assumptions on the response variables: (1) The random errors are independent. (2) The random errors are normally distributed. (3) The random errors have constant variance. (4) The random errors have zero mean.

(continued)

Table 5.2 (continued)

Performance metric	Application
Miscellaneous other test statistics	Breusch–Godfrey test
	Breusch–Pagan test
	Cook's distance
	DFFITS
	Goldfeld–Quandt test
	Leverage
	Park test
	Partial leverage
	Partial regression plot
	Partial residual plot
	Portmanteau test
	PRESS statistic
	Ramsey RESET test
	Regression model validation
	Variance inflation factor
	White test

5.4   A Comparative Example

Commercial software for PLS path analysis and covariance structure methods tends to be inconsistent in its application of algorithms and reporting standards. Thus any comparison of methodologies needs to be standardized in two sets of specific computations—(1) the computation of factor weights (on the so-called outer model), which determine realized values of the associated latent variable, and (2) the reported path coefficients between latent variables (the so-called inner model).

Path analysis software usually allows preselection of the factors that comprise particular latent variable (i.e., reflective links). The weights assigned are most often the factor weights of the first principal component. This is the approach applied here.

Reported path coefficients between latent variables can be 0 dimensional (correlations), 1 dimensional (regression coefficients), or 2 dimensional (covariances). Wright's original path analysis reported correlations (dimensionless), which Wright argued were easy to interpret, and less likely lead to erroneous conclusions.

Tukey promoted regression coefficients (1 dimensional) on the path because they provide information on scale as well as strength of the link. Such path coefficients are typically computed through a sequence of pairwise latent variable OLS regressions (following (Lohmöller, 1988, 1989)). Some software allow alternatives, including PLSR, though with little difference in results. PLSR provides new insights only where the study involves a large, multicollinear dataset, as in spectroscopy, chemometrics, and some other natural sciences. For most practical purposes, regression coefficients will be identical for OLS, PCR, and PLSR applied piecewise to

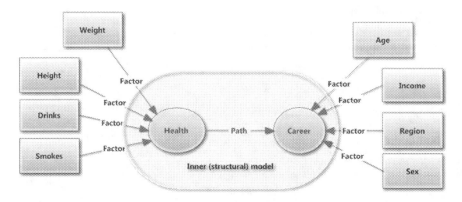

Fig. 5.2 A structural model

path regression, and what differences exist are overwhelmed by the effects of resampling.

Covariance methods naturally generate covariances (2 dimensional) though these are difficult to interpret, and are overly sensitive to the scale of interaction. Because of this, software packages like LISREL and AMOS standardize path coefficients to dimensionless correlations (following Wright) or offer alternatives that compute regression coefficients.

In a typical study, for example, the first step in SEM path analysis would be to choose latent variables, and the construction of a structural model relating them. Usually this starts with a theory, hypothesis, or hunch about the way a process works in the real-world population that is of interest (Fig. 5.2).

An experimental procedure or survey instrument would construct clusters of measurements—for example, the individual questions in the survey questionnaire—and insure that a number of questions pertain to each latent construct to assure that we have properly measured that construct. During execution of the study we would take M measurements (e.g., test questions with Likert scale responses organized into columns in a data matrix) on N individuals (the rows in the data matrix). The researcher, when constructing the survey, will have clustered questions around L particular (generally unmeasured) concepts—latent variables.

The objective of the study is to make decisions about the behavioral relationships between concepts—the links between latent variables. Since the concepts of Health and Career in the structural modal above are not directly measurable, we would like several measurable factors (weight, etc.) to serve as surrogates for these latent variables. Clearly some measurements are better than other for inferring the behavior of latent variables; for example, we might be able to take detailed measures of sunspots, but this is unlikely to be of use in assessing health or career. The ability of factors to consistently measure an immeasurable concept—a latent variable—can be determined through various statistical tests.

Start with a dataset of observations over a set of subjects. Suppose we collect a large number of measurements (columns of data) for each subject. Let's start with a

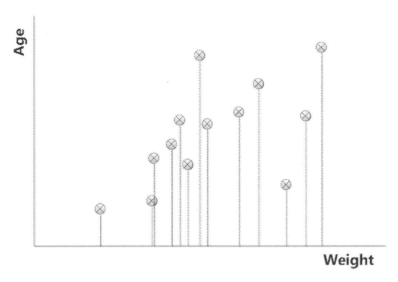

Fig. 5.3 Two-dimensional measurements projected onto a single dimension (Weight, Age) → Weight (the Weight factor line lies on the x-axis)

simplification and assume that there are only two columns of data—call them Age and Weight. Assume that we want to reduce this to one variable which describes the differences in subjects, and let's assume that this is a measured variable—Health. The figure shows how this might be done for a series of measurements on a number subjects, where the 2-dimensional dataset of measurements is reduced to a 1-dimensional dataset of factors. The resulting projected values on the weight axis can be thought of as a 1-dimensional summarization of all of the measurements (Fig. 5.3).

Now this projection is actually not a very good ordination, because large variances in the age values are completely ignored—this does a poor job explaining data variance. Instead we could assume that there is some unobservable (latent) health variable that does a better job of minimizing the variance between our original dataset and the projection. Figure 5.4 depicts one such latent health variable.

The Health variable, even though we can't observe it, better represents the dataset than does the projection on Weight in the sense that measurements are closer to the factor line. This is the concept behind varimax rotations in PCA; it essentially chooses a regression line (based on some fit metric) as the latent variable. In PCA, the extraction of principal components amounts to a variance maximizing (varimax) rotation of the original variable space. For example, in a scatterplot we can think of the regression line as the original axis, rotated so that it approximates the regression line. The resulting projected values on the line can be thought of as a 1-dimensional summarization of all of the measurements.

Assume that we are given the following survey results (Table 5.3)

The covariance structure of this dataset looks like this (Table 5.4).

And the correlations look like this (Table 5.5).

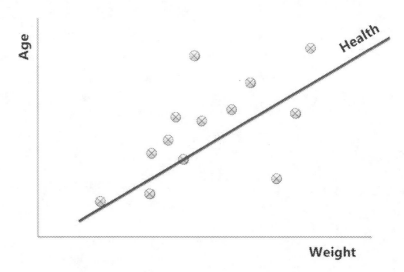

Fig. 5.4 The "Health" latent variable

Table 5.3 An example survey dataset

	Age	Weight	Income	Region	Sex	Drinks	Smokes	Height
Subject #1	36	171	12,100	2	0	0	0	73
Subject #2	40	105	18,600	0	0	0	1	79
Subject #3	33	140	16,900	3	1	1	0	55
Subject #4	34	186	18,100	4	0	1	1	54
Subject #5	22	142	11,000	5	1	1	1	81
...	...	...	...	...	...	...	...	...

Table 5.4 Covariance structure of measured (indicator) variables

	Age	Weight	Income	Region	Sex	Drinks	Smokes	Height
Age	1.00	27.69	1.64	0.20	18.11	24.53	19.12	28.09
Weight	27.69	1.00	6.88	31.40	22.77	7.74	25.33	23.18
Income	1.64	6.88	1.00	13.62	17.91	21.53	27.93	23.48
Region	0.20	31.40	13.62	1.00	18.60	28.39	6.63	10.74
Sex	18.11	22.77	17.91	18.60	1.00	26.55	30.79	3.73
Drinks	24.53	7.74	21.53	28.39	26.55	1.00	16.97	19.70
Smokes	19.12	25.33	27.93	6.63	30.79	16.97	1.00	11.23
Height	28.09	23.18	23.48	10.74	3.73	19.70	11.23	1.00

Table 5.5 Correlations

	Age	Weight	Income	Region	Sex	Drinks	Smokes	Height
Age	1.00	0.87	0.05	0.01	0.57	0.77	0.60	0.88
Weight	0.87	1.00	0.21	0.98	0.71	0.24	0.79	0.72
Income	0.05	0.21	1.00	0.43	0.56	0.67	0.87	0.73
Region	0.01	0.98	0.43	1.00	0.58	0.89	0.21	0.34
Sex	0.57	0.71	0.56	0.58	1.00	0.83	0.96	0.12
Drinks	0.77	0.24	0.67	0.89	0.83	1.00	0.53	0.62
Smokes	0.60	0.79	0.87	0.21	0.96	0.53	1.00	0.35
Height	0.88	0.72	0.73	0.34	0.12	0.62	0.35	1.00

Table 5.6 First five components of the dataset

	Comp. 1	Comp. 2	Comp. 3	Comp. 4	Comp. 5
Age	0.53	−0.14	0.34	−0.11	−0.37
Drinks	0.16	0.46	0.19	−0.43	−0.02
Height	−0.37	0.05	0.61	0.00	0.03
Income	−0.09	0.15	−0.11	−0.72	0.12
Region	−0.37	0.50	0.08	0.35	0.20
Sex	−0.27	0.42	−0.06	−0.08	−0.67
Smokes	0.36	0.40	−0.16	0.00	0.51
Weight	0.28	0.32	−0.40	0.30	−0.31
Component variances	1.86	1.65	1.39	1.25	0.93
Standard deviation	1.36	1.28	1.18	1.12	0.97
Proportion of variance explained	0.21	0.18	0.15	0.14	0.10
Cumulative variance explained	0.21	0.39	0.54	0.68	0.79

Clearly the columns (age, weight, etc.) are not independent; we might expect higher age and drinking to be associated with higher weight, for example. More likely, our research would be motivated by questions about particular abstract concepts such as how a subject's health relates to his or her career choices. Hopefully this had also motivated the particular data collected in the dataset. We can't determine each subject's health or career choices directly, since we didn't measure those factors. But we could infer that certain subsets of the data that we actually did collect (age, weight, etc.) are indicators of health and career choice.

To see if this is a reasonable assumption, we could see whether natural clusters occur in the data that we have collected—this is the process of ordination. Principal component analysis builds a sequence of weighted linear combinations of column data (called components) that explain the variance in the dataset. Component 1 explains the most variance, component 2 the second most variance, and so forth until the column space is spanned (there is no need to understand details of the computation, as computer programs will apply the algorithms) (Table 5.6).

Fig. 5.5 A scree plot

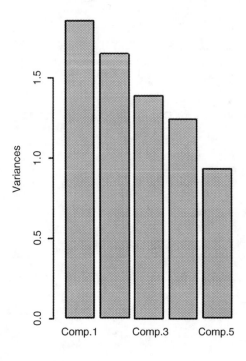

Table 5.7 The first
component (a synthetic latent
variable, with factor weights
for formative indicator links)

Age	0.53
Drinks	0.16
Height	−0.37
Income	−0.09
Region	−0.37
Sex	−0.27
Smokes	0.36
Weight	0.28

A common question arising in ordination is "How many dimensions should I ultimately choose for my model?" Answers are inherently ad hoc—a part of the artistry involved in constructing the research model, hypotheses, and research questions. Researchers often print a scree plot (scree is the gravel that rolls down a slope) tracing out the declining variances of each component. The data can be considered to be optimal for answering research questions about the components with variance greater than 1.0 (in this example, the first four components). This concept is called the Kaiser criterion after Henry Kaiser (1958, 1960, 1974) (Fig. 5.5).

Unfortunately, PCA components may not correspond to constructs of interest in the research; they may even be difficult to relate to any real-world concept. Consider the first component in our sample data, which explains 21 % of variance (Table 5.7).

Table 5.8 The second component (a synthetic latent variable, with factor weights for formative indicator links)	Age	**−0.14**
	Drinks	0.46
	Height	0.05
	Income	0.15
	Region	0.50
	Sex	0.42
	Smokes	0.40
	Weight	0.32

Component 1 appears to reflect some concept affected by age, drinking, smoking, and weight. Let's call this concept "health." We haven't directly measured "health," so we say this is a "latent" or unobserved variable. Latent constructs are often the focus of research; a main motivation of SEM is the ability to measure unobserved latent constructs. These can be especially important in the social sciences where many of the most important subjects involve unobservables (e.g., happiness, tiredness) (Table 5.8)

Component 2 appears to reflect some concept affected by drinking, weight, region, and sex. Let's call this concept "career choice" which is again a latent or unobserved construct. The relationship of factors and their loadings (the numbers to the right of each factor) to our latent concept is more tenuous than with component 1, and indeed we may need to dig down to component 3 to fully understand career choice. But let's suppose that our research question has already posed a question based on two variables—career choice and health. The links between latent and indicator variables thus constructed are sometimes called formative links (as opposed to reflective links which are presumably built into the survey instrument or experiment).

Now consider our two latent variable models of the impact of health on career. Health is a linear function of the observed variables (indicators) of weight, height, drinks, and smokes. The first component of a PCA on these indicators provides factor loadings on the formative links (formative links implying that the linkages were chosen by PCA rather than reflecting the experimental setup).

PCA factor loadings and indicators:

1. Health$=-0.06664$ Drinks-0.72339 Height$+0.179643$ Smokes$+0.663318$ Weight
2. Career$=-0.37341$ Age$+0.494232$ Income-0.72344 Region-0.30484 Sex

Then the four approaches to path analysis discussed to this point would compute the following path coefficient for the one path between the health and career latent variables (Table 5.9).

Note that the arrow in the path diagram represents different things to different methods. In PLS path analysis they serve as a directive to the software for the sequence in which the pairwise regression coefficients are computed by the software. In systems of equations regressions they define the independent and response variables. For correlation and analysis of variance, they can be ignored.

Table 5.9   Path coefficients

	Health-career
Covariance (Jöreskog)	3778.6770
Correlation (Wright, Lohmöller)	0.008797405
Regression (Tukey, Lohmöller)	3.4090
Systems of regressions (OLS)	n/a

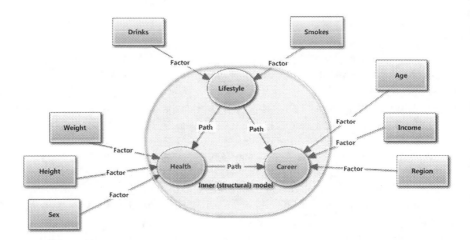

Fig. 5.6   The impact of lifestyle and health on career

Now rearrange the indicators to estimate three latent variables. Health is a linear function of the observed variables (indicators) of weight, height, and sex; career of age, income, and region; and lifestyle of drinking and smoking habits. The first component of a PCA on these indicators provides factor loadings on the formative links (formative links implying that the linkages were chosen by PCA rather than reflecting the experimental setup).

PCA factor loadings and indicators:

1. Lifestyle = −0.70711 Drinks + 0.707107 Smokes
2. Career = 0.32714 Age − 0.60111 Income + 0.729145 Region
3. Health = −0.53761 Height + 0.435841 Sex + 0.721815 Weight (Fig. 5.6) (Table 5.10)

We now have two separate paths: (Career ~ Health + Lifestyle) and (Health ~ Lifestyle). We can apply the systems of equation approaches discussed in this chapter.

Note that the path coefficients on the Health ~ Career path have altered only slightly from the previous model, and that the systems of equation coefficients are close (at least in this model) to the independently OLS regressed paths in Lohmöller's PLS path analysis algorithm. In other situations, there may be significant divergence

Table 5.10 Path coefficients

	Lifestyle-health	Health-career	Lifestyle-career
Covariance (Jöreskog)	−0.6434	−6159.6700	−48.7584
Correlation (Wright, Lohmöller)	−0.0362	−0.0110	−0.0063
Regression (Tukey, Lohmöller)	−2.6038	−4.8190	−197.3000
Systems of equations (OLS)	−2.6038	−4.9243	−210.1337

in the results—especially with multicollinearity and stronger dependencies between latent variables. The values provided by path coefficients vary widely between methods. Wright (1960) argued that correlations, because of their simplicity, gave the most intuitive feel for the relationships between variables. Correlations are dimensionless; regression coefficients provide a sense of linear scale. The covariances in Jöreskog's approach tend to be harder to interpret, as they are inherently nonlinear second moments. But covariance approaches have an advantage of many useful goodness-of-fit metrics. Systems of equation approaches produce the linearly scaled coefficients of PLS path analysis while allowing model fit assessments that are even more extensive than LISREL, AMOS, and other covariance approaches (at a cost of greater complexity).

5.5 A Model of Innovation, Structure, and Governance with Latent Variables in Systems of Regression Equations

This section looks at a real-world application of a systems of regression equation approach used where the model's main constructs are unobservable "latent" variables. The following data was gathered in a survey of innovation, structure, and governance in 198 industrial firms in Northeastern China (Table 5.11).

We might initially propose a five-latent-variable structural model. This would be equivalent to the so-called inner model of a PLS-PA or LISREL approach (Fig. 5.7).

A review of the scree plot of principal components of the data collected revealed that the first five components explained most of the variance in the data (Fig. 5.8).

"Scree" refers to that loose gravel that you slip on when climbing a hill, the plot likened to falling down such a hill. Kaiser proposed a scree plot cutoff criterion that principle components are significant if there is a greater than random likelihood (variance greater than one) of them describing a real-world cluster of traits related to a latent construct (Kaiser, 1958, 1960, 1974; Tabachnick & Fidell, 2001). Here there are five components with variance significantly above one, and five latent variables for the model proposed prior to data collection. But we still need to review the factor loadings to see whether these components correspond to the indicators that the researcher grouped a priori into the five latent variables in the

Table 5.11 Innovation dataset with latent variable choices

Latent constructs	Variable	Mean	Standard deviation	Minimum	Maximum
Governance	Board_Ave_Age	50.66	7.90	24	69
	Board_independence	3.17	0.75	0	6
	Board_size	8.95	2.33	1	19
Financial	Debt_Assets	0.037	0.114	0.000	0.639
	EPS	0.557	0.507	0.00	5.410
	RandD	14,568,840	29,678,910	0.00	182,900,100
	Stock_Price_12.31	19.18	10.61	4.75	68.78
Networking	Political_net	0.641	1.056	0.000	4.000
	Technical_net	1.071	1.489	0.000	4.000
	Industry_net	0.727	1.265	0.000	4.000
Scale	No_Employees	2,749.22	12,653.03	148	177,624
	No_Sales_force	234.88	531.27	0	5,706
	No_Tech_staff	357.47	1,133.08	7	15,733.00
Innovativeness	Patents	22.86	95.17	0	1,240
	New_products	5.914	21.101	0	192

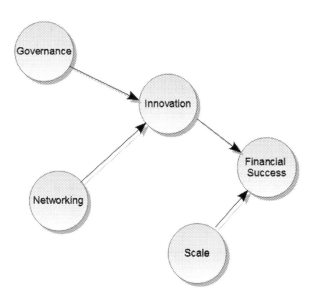

Fig. 5.7 Proposed research path model

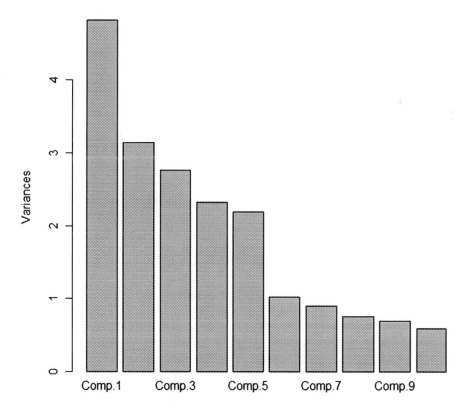

Fig. 5.8 Scree plot from innovation dataset

Table 5.12 Principal components and suggested formative links (factor loadings >0.1 in bold)

Indicator variables	Comp. 1	Comp. 2	Comp. 3	Comp. 4	Comp. 5
Board_Ave_Age	0.005	**−0.234**	**−0.293**	**−0.101**	0.014
Board_independence	0.011	−0.033	**−0.424**	**−0.267**	−0.123
Board_size	−0.043	−0.021	**−0.439**	**−0.142**	−0.107
Debt_Assets	**0.199**	−0.070	0.000	0.072	0.207
EPS	0.022	0.022	**0.285**	**−0.338**	−0.251
Industry_net	0.018	**0.476**	−0.099	0.095	0.068
New_products	**0.354**	0.013	−0.032	0.033	−0.132
No_Employees	**0.433**	−0.058	0.003	0.045	−0.019
No_Sales_force	**0.231**	**0.152**	−0.068	**−0.251**	0.052
No_Tech_staff	**0.430**	−0.064	0.011	0.047	−0.001
Patents	**0.436**	−0.039	−0.008	0.061	−0.060
Political_net	−0.008	**0.331**	−0.082	**0.125**	−0.282
RandD	0.030	**0.181**	−0.058	**−0.193**	0.327
Stock_Price_12.31	0.051	**0.190**	**0.241**	**−0.393**	−0.130
Technical_net	0.053	**0.428**	−0.113	0.113	0.068
Standard deviation	**2.196**	**1.773**	**1.663**	**1.524**	**1.479**
Proportion of variance	**0.230**	**0.150**	**0.132**	**0.111**	**0.104**
Cumulative proportion	**0.230**	**0.379**	**0.511**	**0.622**	**0.726**

research model—governance, networking, innovation, scale, and financial success (Table 5.12).

The first four components explain 62 % of variance; the first three explain 51 % of variance, or over half. If we look at indicators with factor loadings over 0.10, we see that the first three components naturally split into latent variables, based on factor weights. If we add the fourth principal component, things become messier, because of multiple "common factors" in this natural split.

In the first three components, there are only two common factors: (1) the average age of board members and (2) stock price. We could hazard causal explanations. With age comes experience (both good and bad) and older boards tend to be activist in ways that manifest themselves across all aspects of the firm operations—across all three latent variables. And stock price simply reflects success or failure across the firm's operations. So these indicators might best be removed completely from the model, since the information they add is confounding rather than revealing. Note that this principal components approach to constructing latent variables presumes "formative" links—latent variables formed from the principal components of the indicators.

"Reflective" links would reflect the prior research questions and modeling of the researcher. We could have constructed reflective links for this model by taking our original groupings—governance, networking, innovation, scale, and financial success—and computing the first principal component for each. This will impose my prejudices onto my groupings of indicators, which reflect the thinking that presumably went into the survey instrument prior to collecting this data. There may be

Table 5.13 First principal components of researcher-selected latent variables and reflective link factor loadings

Finance	
Debt_Assets	−0.02829
EPS	0.677931
RandD	0.156519
Stock_Price_12.31	0.717712
Scale	
No_Emp	0.096613
No_Sales	−0.70419
No_Tech	−0.70341
Networking	
Industry_net	−0.64955
Political_net	−0.49784
Tech_net	−0.57467
Governance	
Board_Age	−0.38817
Board_Ind	−0.64909
Board_Size	−0.65422
Innovativeness	
New_Products	0.707107
Patents	0.707107

Table 5.14 Correlation matrix of research-selected latent variables

	Financial	Scale	Innov	Net	Gov
Financial	4,645,316	−0.04	0.02	−0.15	−0.01
Scale	−0.04	1,002	−0.83	0.06	0.00
Innovation	0.02	−0.83	79.86	−0.03	0.03
Networking	−0.15	0.06	−0.03	1.79	−0.14
Governance	−0.01	0.00	0.03	−0.14	3.52

good reasons for doing this—we may have more faith in our motivating theory than in our data—but we will pay a price for these prejudices with results that explain a smaller portion of the variance in the data. In this case the factor loadings (first principal components) for the reflective links are given in Table 5.13.

The correlation matrix for the resulting latent variables is given in Table 5.14 (standard deviations along the diagonal). These are the canonical correlation path coefficients between each of the ten possible links between the pairs of latent variables.

Figure 5.9 shows that only three of the ten possible paths are significant: (1) Scale to Innovation; (2) Governance to Networking; and (3) Networking to Financial Success. This suggests that our original model may have been misspecified, since the paths initially postulated were (1) Networking to Innovation; (2) Governance to Innovation; (3) Innovation to Financial Success; and (4) Scale to Financial Success.

The correlation matrix, then, supports a model which looks more like Fig. 5.10 than the original model postulated in the research.

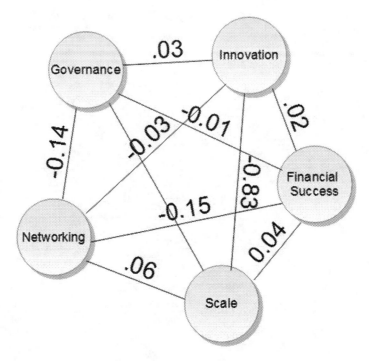

Fig. 5.9 Path coefficients for reflective links

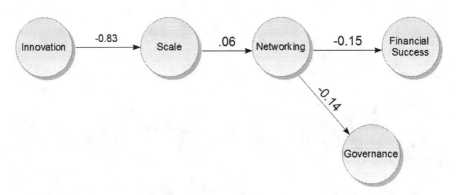

Fig. 5.10 Revised path model suggested by correlation matrix

The prior chapters discussed alternative values for path coefficients, with three main alternatives—correlation, regression, and covariance (Table 5.15).

PCA on the indicators can provide insights into alternative groupings of indicators to link to latent variables (at least from the standpoint of maximizing the variance explained in the dataset) (Table 5.16).

Table 5.15 Alternative choices for path coefficients

	Path			
	Innovation-scale	Scale-net	Networking-financial	Networking-governance
Correlation	−0.83	0.06	−0.15	−0.14
Covariance	−66,127	109	−1,227,056	−0.86
Regression	−0.07	34.45	0.00	−0.07

Table 5.16 Formative link alternatives

Innovativeness indicators	Factor loading
Debt_Assets	0.199
New_products	0.354
No_Employees	0.433
No_Sales_force	0.231
No_Tech_staff	0.430
Patents	0.436
Networking indicators	Factor loadings
Industry_net	0.476
Political_net	0.331
RandD	0.181
Technical_net	0.428
Governance indicators	Factor loadings
Board_Ave_Age	−0.293
Board_independence	−0.424
Board_size	−0.439
EPS	0.285
Stock_Price_12.31	0.241

These are the natural clusters of indicators around latent variables, and explain about half of the variance in the dataset. Unfortunately, they may be less than satisfactory for use in a path model, as path coefficients are likely to be small, since most of the variance in the dataset has already been explained in these clusters. Additionally, they may fail to reflect the researcher's prior beliefs or knowledge about the context in which the research is being conducted. They should be seen as a general guide to the improvement of the structural model.

References

Anderson, C. A. (1983). The causal structure of situations: The generation of plausible causal attributions as a function of type of event situation. *Journal of Experimental Social Psychology, 19*(2), 185–203.

Anderson, J. C., & Gerbing, D. W. (1988). Structural equation modeling in practice: A review and recommended two-step approach. *Psychological Bulletin, 103*(3), 411.

Anderson, T. W., & Rubin, H. (1949). Estimation of the parameters of a single equation in a complete system of stochastic equations. *Annals of Mathematical Statistics, 20*(1), 46–63.

Anderson, T. W., & Rubin, H. (1950). The asymptotic properties of estimates of the parameters of a single equation in a complete system of stochastic equations. *Annals of Mathematical Statistics, 21*, 570–582.

Basmann, R. L. (1957). A generalized classical method of linear estimation of coefficients in a structural equation. *Econometrica, 25*, 77–83.

Basmann, R. L. (1963). The causal interpretation of non-triangular systems of economic relations. *Econometrica, 31*, 439–448.

Christ, C. F. (1994). The Cowles Commission's contributions to econometrics at Chicago, 1939–1955. *Journal of Economic Literature, 32*, 30–59.

Cowles, A. (1933). Can stock market forecasters forecast? *Econometrica, 1*, 309–324.

Cowles, A., III, & Chapman, E. N. (1935). A statistical study of climate in relation to pulmonary tuberculosis. *Journal of the American Statistical Association, 30*(191a), 517–536.

Greene, W. H., & Zhang, C. (2003). *Econometric analysis* (Vol. 5). Upper Saddle River, NJ: Prentice Hall.

Grier, D. A. (2013). *When computers were human*. Princeton, NJ: Princeton University Press.

Kaiser, H. F. (1958). The varimax criterion for analytic rotation in factor analysis. *Psychometrika, 23*(3), 187–200.

Kaiser, H. F. (1960). The application of electronic computers to factor analysis. *Educational and Psychological Measurement, 20*, 141–151.

Kaiser, H. F. (1974). An index of factorial simplicity. *Psychometrika, 39*(1), 31–36.

Koopmans, T. C. (1951). Analysis of production as an efficient combination of activities. *Activity Analysis of Production and Allocation, 13*, 33–37.

Koopmans, T. C. (1957). *Three essays on the state of economic science* (Vol. 21). New York, NY: McGraw-Hill.

Lohmöller, J.-B. (1988). The PLS program system: Latent variables path analysis with partial least squares estimation. *Multivariate Behavioral Research, 23*(1), 125–127.

Lohmöller, J.-B. (1989). *Latent variable path modeling with partial least squares*. Heidelberg, Germany: Physica.

Tabachnick, B. G., & Fidell, L. S. (2001). *Using multivariate statistics*. Needham Heights, MA: Allyn and Bacon.

Theil, H. (1953). *Repeated least squares applied to complete equation systems*. The Hague, Holland: Central planning bureau.

Tukey, J. W. (1954). Causation, regression, and path analysis. In O. Kempthorne et al. (Eds.), *Statistics and mathematics in biology* (pp. 35–66). Ames, IA: Iowa State College Press.

Zellner, A. (1962). An efficient method of estimating seemingly unrelated regressions and tests for aggregation bias. *Journal of the American Statistical Association, 57*, 348–368.

Zellner, A., & Theil, H. (1962). Three-stage least squares: Simultaneous estimation of simultaneous equations. *Econometrica, 30*, 54–78.

Chapter 6
Data Collection, Control, and Sample Size

6.1 The Role of Data

Many questions in social sciences can only be addressed through individual perceptions, impressions, and judgments. A consumer's willingness to pay for a product or
service is a noisy signal, and the consumer has no obligation to follow through on a
purchase intent, no matter how much the researcher might like to infer that "intention" is "action." Such inherently unobservable constructs need to be modeled as a
latent variable. Personal statements of intent, whether they are for purchases, good
deeds, or other promises, can only be considered rough indicators; researchers like
them because they are cheap and easy to collect by questioning the individual. But
like confessions and New Year's resolutions, intentions are pliable and yielding, and
often mendacious. Psychologists have created improved polygraph protocols
involving such questions over nearly a century; yet polygraph evidence is still not
admissible in court. Obtaining truthful and accurate data from surveys and questionnaires is challenging and the quality of information is invariably lacking. Latent
constructs that are of actual interest—ones that help us build theory—are often
unobservable. The only way to understand them is through objective measurement
of related constructs—the indicator variables.

Social science data—particularly financial and economic data that are ultimately
based on double-entry bookkeeping—are often highly multicollinear. Different data
variables tend not to tell us much that is new about either observed or latent constructs. Double entry means that by definition, accounting data counts single data
times or events multiple times. For example, a single sale will appear as a debit to
accounts receivable and credit to sales; later a debit to cash and credit to accounts
receivable—one piece of information (the sales event)—is turned into four data
items. This implies that to get unambiguous statistical results, we need to sample
large numbers of variables, and acquire large datasets of their measurements.

The social sciences are at a disadvantage in data collection when compared with
the natural sciences. Historical records like financial statements and surveys of

© Springer International Publishing Switzerland 2015
J.C. Westland, *Structural Equation Models*, Studies in Systems,
Decision and Control 22, DOI 10.1007/978-3-319-16507-3_6

individual behavior tend not only to be subjective, but they are also one-shot, non-replicable measurements. Except in very contrived situations, social scientists find it difficult to set up a controlled laboratory experiment, and rerun it thousands or millions of times. Of all of the social scientists, econometricians are perhaps the most fortunate with respect to datasets. They inherent the masses of data generated by individual, corporate, and national accounting—in the USA alone, the costs of economic data collection and collation now exceed $1 trillion annually. No other social science comes even close to this level of expenditure on data.

Fisher (1935) described statistics as the study of populations, variation, and methods of data reduction. Samples need to be reduced to summarize information about the population. The three fundamental tasks of the statistician are to:

1. Define the population
2. Identify the sources of variation
3. Decide how the data should be reduced (simplified as a small number of summary statistics)

The use of randomness in some form allows statisticians to use *probability* theory—the branch of mathematics that analyzes random phenomena; contrasted with *statistics* that is the discipline of *inferring* the true state of a population given limited information.

One of the fundamental distinctions made in designing statistical studies is whether these will be *randomized*, or *observational* studies. In survey sampling one often sees "convenience samples" (in the accounting profession these are called "judgmental samples"), implying that random selection procedures have not been used in acquiring the data. The problem is that such samples are not representative of the population (they are only representative of themselves) and the researcher cannot reliably infer anything about the population from the sample. A probability model is required to draw valid inferences from any statistical methods (including SEM methods)—otherwise the conclusions only apply to the data items in the sample.

Observational data may be used to help specify a model (often done in pretest) but is not valid for inferences about the population as a whole. In these situations, precision is less of an issue than bias—the suspicion that researchers have "cooked the books" to support their prejudices. Despite this, the analysis of nonexperimental or quasi-experimental data has captivated statisticians since the field's inception in the seventeenth century. A useful summary of approaches can be found in Copas and Li (1997).

6.2 The Ancient Roots of Model-Data Duality

It is not immediately obvious that data should be distinct and independent from models, no matter how much modern science may be predicated on that assumption. In practice, the human brain's neural networks, programming languages like LISP and APL, Excel spreadsheets, fractals, cellular automata (Wolfram, 2002), and numerous other representations systematically conflate models and data. The control and clarity demanded of scientific argumentation bias research towards a clear model-data dichotomy. This may now be changing with the nascence of computer-intensive analytical tools like the methods surveyed for social network analytics in the final chapter of this book.

The model-dataset duality is ancient, most famously articulated in Plato's *Theory of Forms* (also known as the *Theory of Ideas*, or as Aristotle's *hylomorphism*). It has been a central tenet of intellectual inquiry throughout history. Plato asserted that abstract (but substantial) ideas, and not the material world of change known through measurement, represented the most fundamental kind of reality. Models are representations of reality that are not directly measurable. Structural equation models reflect this duality explicitly—the structural (inner) model is truly an unmeasurable *Platonic form*; the measurement (outer) model contains the actual measurements from the material world.

In The Republic, Plato (in a dialog with his teacher Socrates) presents "the allegory of the cave"—a dialog that captures the essence of the statistical challenge. In the dialogue, Socrates describes a group of people who have lived chained to the wall of a cave all of their lives, facing a blank wall. The people watch shadows projected on the wall by things passing in front of a fire behind them, and begin to ascribe forms to these shadows. Socrates saw philosophers—the scientists of those days—as "freed" prisoners who can use their mental tools to perceive reality rather than the shadows. And like the residents of Plato's cave, who know the world only through shadows, researchers cannot completely know the "true" state of nature—the reality. Rather they have to make do with artificial measurements (shadows lacking dimension and color, and which change shape depending on the angle from which you measure them). Research can only hope to collect a sufficient number of measurements over time to gain some insight into this unseen reality. These measurements are collectively termed data. But they can be unreliable shadows of the things they represent.

The hypothetico-deductive model arose in its modern form with the experiments of Galileo Galilei in the sixteenth century. Growing more complex and reliable over the past five centuries, it has so far managed to fend off competitors in scientific discourse, and today the hypothetico-deductive model remains perhaps the best understood theory of scientific method.

The hypothetico-deductive model for scientific inquiry proceeds by formulating a hypothesis in a form that could conceivably be falsified by a test on observable data. This is typically accomplished through the following steps:

1. Specify a model of the real world based on hunches, experience, exploratory data search, and other subjective means.
2. Segment the research model into a series of "yes/no" questions called hypotheses.
3. Conjecture predictions from the hypothesis. Identify the expected characteristics of real-world observations (i.e., the data) that you would expect to find were the hypothesis true.
4. Test (i.e., experiment) by collecting evidence (i.e., data) germane to each hypothesis and then applying the tools of inference (i.e., statistics) to draw conclusions.

5. Based on strength, cost, and feasibility of additional data collection, revise and improve the experiment until the desired level of certainty about the model truth or falsehood is achieved.

Today, the language of the hypothetico-deductive model is central to the social sciences. But in contrast to the natural science, social science data measurements can be ephemeral and ethereal. Consider, for example, a survey of people's intentions to do something—for example a New Year's resolution made on January 1st. Assume that you asked 100 subjects to rank on a scale of 1 (intend to lose 0 lb) to 7 (intend to lose 20 lb) their intention to lose 20 pounds. Now suppose that you used the result of this survey—sample mean $\bar{x} = 3.5$ and standard deviation of $s = 1$—to predict how many pounds the subjects would lose. You assume that intentions are distributed normally (even though they are integers that only run from 1 to 7). You predict that the average subject will lose 10 lb. Do you think that would be an accurate prediction of what subjects *actually* had done by March 31st? Indeed, given that this is a New Year's resolution, it would not be surprising if a substantial number of subjects had added on weight at the end of 3 months, or had completely given up any intention of losing weight by that time (the survey measurements make no accommodation for either the intention or reality of adding weight).

Our example suggests that at least two questions are pivotal in dataset choices:

(1) Is it representative of the "population" that is assumed in the model?
(2) How much will the collected data cost as a function of quality and quantity?

The first question—"Is it representative of the 'population' that is assumed in the model?"—requires a strategy for data acquisition that is tied to the specific factors or components in the model. Ideally these model factors should be the columns of your dataset—but you usually are not that lucky. In the context of this book, structural equation models give you the luxury of an inner structural model that contains the factors important in the research question, but then allows the collection of data that does not directly measure these structural factors. This ends up being important when models contain many abstract concepts, as is common in the social sciences.

Still, the data needs to say something about the abstract factors in the model. This is where the representativeness of the dataset becomes important. Model abstractions make sweeping generalizations about, for example, teenagers, consumers, voters, or other such groupings of people. Let's assume that we have a question about teenagers. The total population of teenagers (ages 13–19) in the USA is about 30 million, simply too large for an affordable dataset. The solution is to sample a representative subset of the population, and extrapolate to the population. But how do we assure that the sample is representative? One approach is to assure that each and every one of those 30 million teenagers has an equal probability of appearing in the sample (much harder than it might first seem, which is the problem the Census Bureau faces every 10 years). This is achieved through random sampling strategies. Random sampling avoids selection biases—for example choosing only 19-year-old college sophomores because it is easy to find them in classrooms.

In laboratory or agricultural tests that involve testing the effect of treatments, even more involved designs may be concocted to assure representativeness. Experimental design is a separate discipline in statistics that dictates information-gathering exercises where variation is present, whether under the full control of the experimenter or not. The analysis of variation is central to SEM analysis, and the degree to which researchers can control this dictates methodology.

Laboratory experiments attempt to cede as much control as possible over treatments and effects to the experimenter. In contrast, natural or quasi-experiments may contain many of the features of laboratory experiments, but are one-shot, non-repeatable experiments. Almost all social science must make do with natural experiments—naturally occurring instances of observable phenomena that approach or duplicate a scientific experiment. In contrast to laboratory experiments, these events aren't created by scientists, but yield scientific data. Natural experiments are a common research tool in fields where artificial experimentation is difficult, such as business, cosmology, epidemiology, and sociology. For example consider the spread of early humans across the Pacific Ocean—and important area of research for historians. The distribution of populations between islands was essentially random, allowing researchers to treat different groups as independent societies drawn from a common pool. Hypotheses could then be tested in different contexts without fear that an unobserved factor is the cause of differences between island populations.

The second question—"How much will the collected data cost as a function of quality and quantity?"—requires that we understand the marginal cost of each data item collected. This marginal cost tends to have a substantial impact on cash-strapped academics, and dictates experimental design and scope. At the extremes—for example large telescopes, particle accelerators, and space flights—data collection may come only once in a lifetime, and a limited quantity is available during the research window.

Even without such extreme constraints, there is pressure to economize, perhaps even to cheat, through a variety of shortcuts:

1. Select data that is convenient (college sophomores) rather than representative (random).
2. Double and triple count data already collected, for example, through improper use of bootstrapping.
3. Opportunistically adjust the model to make the dataset seem like it contains more information, through stepwise regression and other techniques to maximize fit statistics without rethinking what the model states about reality.
4. Claim research findings where there exist only compelling patterns, a strategy sometimes seen in data visualization (for example in brain imaging), ordination, simulations, and so forth. Methodologies that are useful for data exploration and model specification, for example factor analysis or partial least squares, are not appropriate for testing hypotheses and determining the goodness of fit of data.

These are all essentially ways of *gaming* the formal methods of science; unfortunately they undermine the credibility of the streams of research in which they are used.

They may use the vernacular of statistics to lend credibility to their conclusions, when in fact their conclusions have been chosen in advance, and the statistics are rigged to favor the preselected outcomes.

6.3   Data: Model Fit

A major problem that arose in the shift from principal component analysis (PCA)-defined latent variables (formative links proposed by Wold) to researcher-defined latent variables (reflective links) is that researchers may get the wrong indicators matched with a particular latent variable—this may generate common factor bias and other biases on the path coefficients. Cronbach's alpha and common factor tests are designed to spot this, but only for a single latent construct.

The larger problem is how the information in the indicators aligns with the research model (the latent constructs and paths). There may be information in the dataset on less than the complete set of model latent constructs. For example, if there are six latent constructs, and only three PCA components with eigenvalues greater than 1, then there is probably only information in the dataset to support three latent constructs—and these may not be exactly the latent constructs chosen by the researcher.

The proper way to determine model-data fit is to run a PCA on the data, use the Kaiser criterion to select significant components (those with eigenvalues >1), and then see which latent constructs they align with. Then either the model needs to be adjusted to fit the data or more data (either more observations or more attributes) needs to be collected.

No matter what methodology is used in analysis of the indicator observations, the performance of the researcher's structural model will be explained in terms of its explanatory power relative to some unconstrained optimal partitioning of the data into latent variables. The standard is commonly set by PCA which selects latent variables (components) to maximize the explanation of variance in the data.

There are actually many ways in which data can be misaligned. Figure 6.1 and Table 6.1 describe the various ways that we may misalign model and data in the designing of reflective links in constructing latent variables.

As the structural model expands to include more and more latent variables, the opportunities for misspecification increase exponentially. Cronbach's alpha and other related measures become more and more difficult to interpret, as there will be an exponentially expanded set of ways that indicators can be misassigned to latent variables.

This is a weakness in statistics like Cronbach's alpha. Such weaknesses can be circumvented by returning to the initial objectives behind the statistic—to resolve the problem in the research's latent constructs, rather than letting a PCA "organically" select latent variables (principal components) to maximize the

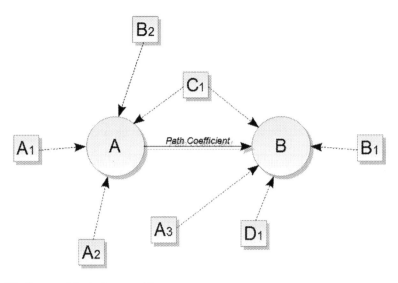

Fig. 6.1 Data-model misalignment biases

Table 6.1 Data-model misalignment biases

		Error bias (+/−)	
		Path coefficient	Variance explained
Indicator	Description		
A_1	*Valid indicator* of latent variable A, factor loading will be large	0	+
A_2	*Valid indicator* of latent variable A, factor loading will be large	0	+
A_3	This should have been modeled as an *indicator of latent variable B*; factor loadings for B will be small and for A will be large	−	−
B_1	Valid indicator of latent variable B, factor loading will be large	0	+
B_2	This should have been modeled as an *indicator of latent variable A*; factor loading for B will be small	+	−
C_1	This is a *common factor* for both latent variables A and B; factor loadings for A and B will be large, even though only one of these may have been included in the model (most path analysis does not allow for shared indicator variables)	+	−
D_1	This is an indicator of some *omitted latent variable*, factor loading will be small or zero. The omitted variable needs to be included in the structural model	+	−

variance explained. The data-model fit problem needs to be resolved by either proposing a new model and theory around the exploratory analysis of the dataset; or alternatively, by choosing a dataset that contains significant information about all of, not just a subset of, the model parameters.

Researchers set out to answer specific research questions that require definition of a set of concepts—both measurable and abstract. Instruments and studies are designed to collect data, which often comprise the majority of research expenditures. Unfortunately, the data do not always neatly offer up answers to the questions asked of them. Data may be incomplete, or answer different questions than asked, or simply provide insufficient information for an answer. This is not really controllable in advance—it is part of the inherent risk in inquiry. So the researcher can usually be assured that the information in the dataset and the information required to answer the research questions will not completely coincide. Either the research questions and hypotheses need to be modified to fit the information in the data or more data needs to be collected. This additional data collection can acquire more indicators (information on latent variables that is missing in the original dataset) or can acquire more observations (allowing more precise estimates, or the identification of smaller effects).

Path modelers sometimes refer to two contrasting ways of defining the links between indicator (also called "manifest") variables and latent variables: reflective and formative (Dijkstra, 1983; Lauro & Vinzi, 2002). Direction of arrows may be used to distinguish reflective from formative links—but this is misleading, as the two definitions are not distinguished by causal direction, rather by the way they cluster around a latent construct. In fact, any assertion of causal direction in indicator-latent variable links would be dubious, since the indicators are observed and the latent variables are unobserved.

Reflective links are model driven—they are created in the design of the research study. In the case of *reflective* indicator links, typical of classical factor analysis models, the latent constructs give rise to observed variables that covary among them and the model aims at accounting for observed variances or covariances. This is typically what is encountered in surveys, where clusters of questions are intended to glean information about a particular unobservable (latent) construct. *The researcher creates a cluster of reflective variable links around a latent variable in the construction of the model and the survey instrument.*

Formative links are data driven—they are inferred *ex posteriori* from data that is already collected. Formative indicator links are emergent constructs—combinations of observed indicators and are not designed to account for observed variables. The components of a typical PCA analysis are such combinations of observed indicators—these components don't necessarily correspond to specific constructs, observed or not. *The researcher creates a cluster of formative variable links around a latent variable through factor analysis or PCA computed with a mathematical objective in mind, e.g., minimizing variance and dimensional reduction.*

Ideally we would like for formative and reflective indicator links to be identical. A successful experimental design will preselect indicator variables for each latent construct that are strongly correlated with each other—which are multicollinear.

This helps validate the assertion that each of several questions on a survey, or measurements in an experiment, represents exactly the same thing—measurements in an unobserved latent construct. In implementation, a latent variable is simply a linear function of the indicators and the latent variable exists in concept only as the realized value of this linear function. This is the reasoning behind the three tools for insuring that the indicator variables are informative, consistent, and appropriate surrogates for the unobserved latent variable:

1. *Harman's single-factor* test and *Cronbach's alpha* are concepts in the validation of reflective links.
2. The related concept of the *Kaiser criterion* is used in the choice of indicators in formative links.

The basic concepts are straightforward and simple. Common factor bias is the same as common method variance—"factor" means the same thing as "indicator," and "method" is the way in which this indicator is chosen. Bias and variance are used similarly as well, to indicate the implied changes in the latent variable. Technically, these are portions of the model variation that can be ascribed to problems in the measurement method (e.g., the way in which the survey instrument was constructed; whether the questions were "on scale" and "balanced") rather than the (latent) constructs of interest. These are presumed to cause systematic measurement error and bias the estimates of the "true" relationship among theoretical constructs (Koopmans, 1951). In biasing the model path estimates, these can lead to problems in hypothesis tests.

Where confirmation is the objective, one problem that can arise in building the structural model completely without reference to the data (i.e., all reflective indicator links) is that the latent constructs chosen by the researcher may be substantially different than those that would drop out of an exploratory factor analysis. Harman's one-factor test (Koopmans, 1951, 1963; Podsakoff & Organ, 1986) has been suggested as a test for common factor bias or common method variance. The most common reason that this test is needed is that the model is constructed without reference to clustering in the underlying data; that is, it is entirely theory driven (not in itself a bad thing).

Harman's single-factor test performs a factor analysis or PCA on all of the indicators collected for the study (presumably these will all be reflective indicator links when you use this test) and assesses whether the very first factor or component is significantly larger than all of the other components. If so, it is assumed that your unobserved *latent* variables are multicollinear, and there is a "common method bias"—i.e., a single unobserved factor, perhaps introduced in the survey or experimental design indicative of common method variance—that influences all of the latent variables and overstates the correlations between them.

Conversely, one can inspect each latent variable and its associated cluster of indicators to see if these are appropriate choices for reflective indicators. Cronbach's alpha is used as a coefficient of reliability for such choices—it provides a measure of the internal consistency or reliability of a psychometric test score on questionnaires (Koopmans, 1957). It was first named α (alpha) by Lee Cronbach in 1951, as he had

Table 6.2 Guidelines for assessing internal consistency

Value of Cronbach's alpha	Internal consistency
$\alpha \geq 0.9$	Excellent
$0.9 > \alpha \geq 0.8$	Good
$0.8 > \alpha \geq 0.7$	Acceptable
$0.7 > \alpha \geq 0.6$	Questionable
$0.6 > \alpha \geq 0.5$	Poor
$0.5 > \alpha$	Unacceptable

intended to continue with further coefficients. Cronbach's alpha statistic is widely used in the social sciences, business, nursing, and other disciplines. It attempts to answer the extent to which questions, test subjects, indicators, etc. measure the same thing (i.e., latent construct). Nunnally (1967) provides the following guidelines for assessing internal consistency using Cronbach's alpha (Table 6.2).

Henry Kaiser and Lee Cronbach were both on the faculty of the School of Education at University of Illinois in the 1950s. As colleagues they worked together on applications of the Kaiser criterion and Cronbach's alpha and developed interdependent implementation theories (Cronbach & Meehl, 1955; Joreskog, Sorbom, & Magidson, 1979; F. H. Kaiser, 1991; H. Kaiser, 1992; Likert, Roslow, & Murphy, 1934). These discussions have generated a mountain of research papers—especially if one adds the dependent common factor bias stream to it—for what is a very simple concept.

6.4 Latent Variables

Latent variables are in practice constructed from linear combinations of indicator variables. We would like each latent variable to represent a unique and meaningful construct, even though it is unobserved. And we would like to collect sufficient indicator data on each of the latent constructs that we include in the model. Put differently, we would like the data items we have collected to coalesce, under some clustering algorithm, to clusters that match with the latent variables. There are two ways to do this:

1. We can collect the data, run a PCA (or other factor analysis) on that data, and include latent constructs for the most significant components derived from the PCA. The Kaiser criterion would suggest that all components with eigenvalues over 1 be included. Indicator variables and links are *formative* in this case.
2. We can build a theory-based model, and then collect data to test the model (and thus indirectly, the theory). This requires two additional steps over the first approach:

   (a) The experiment or survey instrument needs careful construction around the model parameters. Data scaling, granularity, location, and reliability are all confounding issues in correctly constructing survey instruments.

(b) Once the data is collected, it needs to be tested to make sure that the clusters that actually exist in the data correspond to the expected clustering (i.e., the *reflective* indicator-link constructs for each predefined latent construct). Harman's one-factor test and Cronbach's alpha are generic tests.

In the model-driven case where *reflective* indicator-link constructs are built from theory, the researcher takes on an additional obligation—to assure that data is collected for each latent construct, and that this data is reliable, consistent, and adequate to support the conclusions of the research. Cliff (1988) argues that using both Cronbach's alpha *and* the Kaiser criterion to identify components with significant eigenvalues is required to properly validate the adequacy and reliability of the data. These two related tests need to be used together to assess validity of indicator-latent variable clustering. Though H. Kaiser (1992) argues for slightly differing criteria, it is clear that this expanded notion of principal component testing for data-model fit was on the minds of both Kaiser and Chronbach when they developed their assessments in the 1950s.

Problems of data-model fit—whether you are discussing common factor bias, interfactor reliability, or some other criterion—can be avoided a priori through a pretest of the clustering of indicator data. Common factor bias occurs because procedures that should be a standard part of model specification are in practice left until after the data collection and confirmatory analysis. Jöreskog developed PRELIS for these sorts of pretests and model re-specifications. If this clustering shows that the indicators are providing information on fewer variables than the researchers' latent SEM contains, this is an indication that more indicators need to be collected that will provide (1) additional information about the latent constructs that don't show up in the cluster analysis and (2) additional information to split one exploratory factor into the two or more latent constructs the research needs to complete the hypothesized model. In exploratory factor analysis, the two tests that are most useful for this are the *Kaiser (1960) criterion* that retains factors with eigenvalues greater than 1 (unless a factor extracts at least as much information as the equivalent of one original variable, we drop it) and the *scree test* proposed by Cattell (1966) that compares the difference between two successive eigenvalues and stops taking factors when this drops below a certain level. In either case, the suggested factors are not necessarily the latent factors that the researcher's theory would suggest—rather they are the information that is actually provided in the data, this information being the main justification for the cost of data collection. So in practice, either test would set a maximum number of latent factors in the SEM if that SEM is to be explored with one's own particular dataset.

Common factor bias can be avoided a priori through a pretest of the clustering of indicator data. Common factor bias occurs because procedures that should be a standard part of model specification are in practice left until after the data collection and confirmatory analysis. Jöreskog developed PRELIS for these sorts of pretests and model re-specifications. If this clustering shows that the indicators are providing information on fewer variables than the researchers' latent SEM contains, this is an indication that more indicators need to be collected that will provide (1) additional information about the latent constructs that don't show up in the cluster analysis and (2) additional information to split one exploratory factor into the two or more latent constructs the research needs to complete the hypothesized model. In exploratory factor analysis, the two tests that are most useful for this are the *Kaiser (1960) criterion* that retains factors with eigenvalues greater than 1 (unless a factor extracts at least as much information as the equivalent of one original variable, we drop it) and the *scree test* proposed by Cattell (1966) that compares the difference between two successive eigenvalues and stops taking factors when this drops below a certain level. In either case, the suggested factors are not necessarily the latent factors that the researcher's theory would suggest—rather they are the information that is actually provided in the data, this information being the main justification for the cost of data collection. So in practice, either test would set a maximum number of latent factors in the SEM if that SEM is to be explored with one's own particular dataset.

6.5 Linear Models

All four of the methods presented in this book—correlation, PLS path analysis, covariance structure methods, and systems of equation regression—use linear models and generalizations. These are appropriate where the population characteristics are linear; but they are misleading where they are not. Many real-world relationships are nonlinear—not just a little, but substantially nonlinear: for example, the technology acceleration depicted by Moore's law (computing power doubles every 18 months) is exponential; the value of social networks as a power of the number of members; and the output of a factory has declining returns to scale. It is important to always look for a physical model underlying the data. Assume a linear model as a starting point only or a simplification which may be useful, but which cannot go unexamined.

Models may be conceived and used at three levels (Ford, 2000). The first is a model that fits the data. A test of goodness of fit operates at this level. Linear models often fit the data (within a limited range) but do not explain the data. A second level of usefulness is that the model predicts future observations—it is a *forecast* model. Such a model is often required in screening studies or studies predicting outcomes. Influential arguments by Friedman (1953) suggest that forecast models must form the basis of economic research. A third level is that a model reveals unexpected features of the situation being described—it is a *structural* model.

The term *structural model* is ambiguous. In SEM the structural model is a linear network model of unobservable factors, where the links are canonical (implied) correlations between unseen factors. But the broader definition of structural model embraces nonlinearity and is expected to fully explain the data and the population. This is a tall order, and one that foments arcane debates among statisticians and philosophers. Those of us interested in just completing our data analysis can safely sidestep the debate in most circumstances.

As a measure of such strength, correlation should be large and positive if there is a high probability that large or small values of one variable occur (respectively) in conjunction with large of small values of another; and it should be large and negative if the direction is reversed (Gibbons & Chakraborti, 1992).

6.6   Hypothesis Tests and Data

Classical (Neyman-Pearson) hypothesis testing requires assumptions about underlying distributions from which the data were sampled. Usually the only difference in these assumptions between parametric and nonparametric tests is the assumption of a normal distribution; all other assumptions for the parametric test apply to the nonparametric counterpart as well. In SEM, while PLS path analysis does not make distributional assumptions, the covariance and simultaneous equation regression approaches both often assume normal datasets; nonparametric SEM approaches do not exist.

Hypothesis testing very often assumes that data observations are:

1. Independent
2. Identically distributed (come from the same population with the same variance)
3. Follow a normal distribution

These assumptions are made both for convenience and for tractability; and for simple models they may be good enough. But at a minimum, the researcher is obligated, prior to model fitting, to test the dataset to assure that data are independent, and that they represent the population. This is often accomplished through various exploratory tests, such as histograms of observations. The third assumption tends to be a substantial hurdle in survey research where responses are recorded on a Likert scale (R. Likert, 1932). Likert scale data is discrete and truncated. It is categorical or multinomial, where a normal distribution offers a poor approximation.

6.7   Data Adequacy in SEM

The complex, networked structures of SEM create significant challenges for the determination of sample size and adequacy. From a practical viewpoint, sample size questions can take three forms:

1. *A priori*: what sample size *will be sufficient* given the researcher's *prior beliefs* on what the minimum effect is that the tests will need to detect?
2. *Ex posteriori*: what sample size *should have* been taken in order to detect the minimum effect that the researcher *actually detected* in an existing (either sufficient or insufficient) test? If the *ex posteriori* measured effect is *smaller* than the researcher's prior beliefs about the minimum effect then sample size needs to be increased commensurately.

3. *Sequential test optimal stopping*: this is typically couched in terms of a sequential test optimal stopping context, where the sample size is incremented until it is considered sufficient to stop testing.

In addition, not all sample points are created equal. A single sample data point copied three times over still has only one single sample data point worth of information. Even where complex "bootstrapping" processes are invoked to duplicate sample points, it is doubtful whether new information about a population is actually created (and where it is, it might better be injected into the data through Bayesian methods, or aggregation). If our research question is about the wealth of a consumer group, then a dataset of colors of the sky at different times of the day will not provide information relevant to the research question. Sample data points will contain differing amounts of information germane to any particular research question. Several data points may contain information that overlaps, which is one cause of multicollinearity. The distribution of random data may also differ from modeling assumptions, a problem that commonly occurs in the SEM analysis of Likert scale survey data.

To this day, methodologies for assessing suitable sample size requirements remain a vexing question in SEM-based studies. The number of degrees of freedom consuming information in structural model estimation increases with the number of potential combinations of latent variables, while the information supplied in estimating increases with the number of measured parameters (i.e., indicators) times the number of observations (i.e., the sample size)—both are nonlinear in model parameters. This should imply that requisite sample size is *not* a linear function solely of indicator count, even though such heuristics are widely invoked in justifying SEM sample size. Monte Carlo simulation in this field has lent support to the nonlinearity of sample size requirements, though research to date has not yielded a sample size formula suitable for SEM.

Since the early 1990s, researchers in marketing, MIS and other areas of business, sociology and psychology have alluded to an ad hoc rule of thumb requiring the choosing of ten observations per indicator in setting a lower bound for the adequacy of sample sizes. Justifications for this *rule of 10* appear in several frequently cited publications (Barclay, Higgins, & Thompson, 1995; Chin, 1998; Chin & Newsted, 1999; Kahai & Cooper, 2003) though none of these researchers refers to the original articulation of the rule by Nunnally (1967) who suggested (without providing supporting evidence) that in SEM estimation "a good rule is to have at least ten times as many subjects as variables."

Goodhue, Lewis, and Thompson (2006, 2007) and Goodhue, William, and Thompson (2007) studied the *rule of 10* using Monte Carlo simulation; they found that with "rule of 10" samples that PLS-PA analysis had *inadequate power to detect small or medium effects at small sample*. This finding was fully expected, as similar PLS-PA studies had discredited the "rule of 10" ever since Nunnally's (1967) proposal. Bollen (1989) stated that "though I know of no hard and fast rule, a useful suggestion is to have at least several cases per free parameter" and Bentler (1989) suggested a 5:1 ratio of sample size to number of free parameters. But was this the right question? Typically their parameters were considered to be indicator variables in the model, but unlike early path analysis, structural equation models today are typically estimated in their entirety, and the number of unique entries in the covariance matrix is $\dfrac{p(p+1)}{2}$ when p is the number of indicators. It would be reasonable to assume that the sample size is proportional to $\dfrac{p(p+1)}{2}$ rather than p. Unfortunately, Monte Carlo studies conducted in the 1980s and 1990s showed that the problem is somewhat more subtle and complex than that, and sample size and estimator performance are generally uncorrelated with either $\dfrac{p(p+1)}{2}$ or p.

Difficulties arise because the p indicator variables are used to estimate the k latent variables (the unobserved variables of interest) in SEM, and even though there may be $\dfrac{p(p+1)}{2}$ free parameters, these are not individually the focus of SEM estimation. Rather, free parameters are clustered around a much smaller set of latent variables that are the focus of the estimation (or alternatively, the correlations between these unobserved latent variables are the focus of estimation). Tanaka (1987) argued that sample size should be dependent on the number of estimated parameters (the latent variables and their correlations) rather than on the total number of indicators, a view mirrored in other discussions of minimum sample sizes (Browne & Cudeck, 1989, 1993; Gerbing & Anderson, 1985; Geweke & Singleton, 1980). Velicer and Fava (1998) and Fava and Velicer (1992a, 1992b) went further, after reviewing a variety of such recommendations in the literature, concluding that there was no support for rules positing a minimum sample size as a function of indicators. They showed that for a given sample size, a convergence to proper solutions and goodness of fit were favorably influenced by (1) a greater number of indicators per latent variable and (2) a greater saturation (higher factor loadings).

Marsh and Bailey (1991) concluded that the *ratio of indicators to latent variables* rather than just the number of indicators, as suggested by the rule of 10, is a substantially better basis on which to calculate sample size, reiterating conclusions reached by Boomsma (1982a, 1982b) who suggested using a ratio $r = \dfrac{p}{k}$ of indicators to latent variables. Information input to the SEM estimation increases both with more indicators per latent variable and with more sample observations. A series of studies (Ding, Belicer, & Harlow, 1995) found that the probability of rejecting true models at a significance level of 5 % was close to 5 % for $r = 2$ (where r is the ratio of indicators to latent variables) but rose steadily as r increased— for $r = 6$, rejection rates were 39 % for sample size of 50; 22 % for sample size of 100; 12 % for sample size of 200; and 6 % for sample size of 400.

Boomsma's (1982a, 1982b) simulations suggested that a ratio r of indicators to latent variables of $r = 4$ would require a sample size of at least 100 for adequate analysis, and for $r = 2$ would require a sample size of at least 400. Marsh et al. (Marsh, Balla, & McDonald, 1988; Marsh, Balla, & Hau, 1996; Marsh, Hau, Balla, & Grayson, 1998) ran 35,000 Monte Carlo simulations on LISREL CFA analysis, yielding data that suggested that $r = 3$ would require a sample size of at least 200; $r = 2$ would require a sample size of at least 400; and $r = 12$ would require a sample size of at least 50. Consolidation and summarization of these results suggest sample sizes

$$n \geq 50r^2 - 450r + 1{,}100$$

where r is the ratio of indicators to latent variables. Furthermore, Marsh et al. (1996) recommend $r = 6$ to 10 indicators per latent variable, assuming that 25–50 % of the initial choices add no explanatory power, which they found often to be the case in their studies. They note that this is a substantially larger ratio than found in most SEM studies, which tend to limit themselves to three to four indicators per latent variable. It is possible that a sample size rule of ten observations per indicator may indeed bias researchers towards selecting smaller numbers of indicators per latent variable in order to control the cost of a study or the length of a survey instrument. Figure 6.2 depicts the sample size implied in Boomsma's simulations (Fig. 6.2).

Boomsma's guideline is couched in terms of number of indicator, or measured variables. In an Excel spreadsheet of data, this would imply that the minimum number of rows (sample points) that are required is some function, $50\left(\dfrac{\text{columns}}{\text{latent}}\right)^2 - 450\dfrac{\text{columns}}{\text{latent}} + 1{,}100$, of the number of columns (indicator variables). This is an improvement on the *rule of 10*, which just multiplies the number of columns by a constant. But it is not sufficient for hypothesis testing, because it fails to take into account significance and power of the test, minimum detectable effect, and scaling.

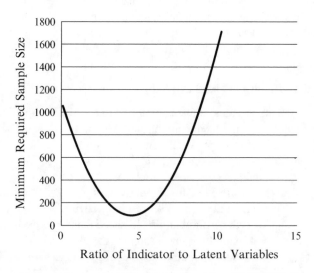

Fig. 6.2 Marsh and Bailey (1991) and Boomsma (1982a, 1982b) calculations of minimum required sample size

6.8 Minimum Sample Size for Structural Equation Model Estimation

This section presents an algorithm for computing the lower bound on sample size required to confirm or reject the existence of a minimum effect in an SEM at given significance and power levels. Where SEM studies are directed towards hypothesis testing for complex models, with some level of significance α and power $1-\beta$, calculating the power requires first specifying the effect size δ you want to detect. Funding agencies, ethics boards, and research review panels frequently request that a researcher perform a power analysis; the argument is that if a study is inadequately powered, there is no point in completing the research. Additionally, in the framework of SEM the assessment of power is affected by the variable information contained in social science data. Table 6.3 summarizes the notation used.

We start by asking "What is the *lower bound on sample size n* for confirmatory testing of SEM as a function of these design parameters?" We want to detect a minimum correlation (effect) δ in estimating k latent (unobserved) variables, at significance and power levels $\left(\alpha^{*}, 1-\beta\right)$. In other words, devise an algorithm $f\left(\cdot\right)$ such that $n = f\left[k, \delta \middle| \alpha^{*}, \beta\right]$.

Table 6.3 Notation for sample size calculations

p	Number of parameters (indicators) in the SEM
k	Number of latent variables in the SEM
n	Computed sample size lower bound
$\left[\tilde{X}, \tilde{Y}\right]$ and $[Xi, Yi]$	Bivariate normal random latent variables (and their realization) in the SEM
$X_{1:n}$ " $X_{2:n}$ ··· $Y_{1:n}$ " $Y_{2:n}$ ···	Order statistics of the (Xi, Yi) sample values; the first index is rank, and the second is sample size
$Y_{[i:n]}$	Concomitant of the ith order statistic; $Y_{[i:n]}$ is the Y sample value associated with the $Xi_{:n}$ sample value in the sample pairs (Xi, Yi).
δ	Minimum effect size that our computed sample size can detect
ρ	Unknown correlation for a bivariate normal random vector $\left[\tilde{X}, \tilde{Y}\right]$
$\hat{\rho}_{G}$	Estimator of Gini correlation ρ_{G}
$\left[\hat{\mu}_{G}; \hat{\sigma}_{G}\right]$	Mean and standard deviation estimators for Gini correlation
$\left[\alpha^{*}; 1-\beta\right]$	Significance and power of test
α	The Šidàk corrected significance for discriminations between possible SEM link combinations at a resolution of δ
$\left[z_{1-\alpha}; z_{1-\beta}\right]$	Rejection bound at significance α and nonrejection bound at power $1-\beta$; we substitute the quantile function (inverse cumulative normal) $\Phi^{-1}\left(x\right)$ for zx in calculations

We adopt the standard targets for our required Type I and II errors under Neyman-Pearson hypothesis testing of $\alpha^* = 0.05$ and $\beta = 0.20$; but these requirements can be relaxed for a more general solution. Structural equation models are characterized here as a collection of pairs of canonically correlated latent variables, and adhere to the standard normalcy assumption on indicator variables. This leads naturally to a deconstruction of the SEM into an overlapping set of bivariate normal distributions. Make the assumption that an arbitrarily selected pair of latent variables, call them X and $\acute{Y}$, are bivariate normal with density function

$$f\left(x,y|\mu_x,\mu_y,\rho,\sigma_x,\sigma_y\right) = \frac{1}{2\pi\sigma_x\sigma_y\sqrt{1-\rho}}$$
$$\exp\left[-\frac{1}{2\left(1-\rho^2\right)}\left(\frac{\left(x-\mu_x\right)^2}{\sigma_x^2} + \frac{\left(y-\mu_y\right)^2}{\sigma_y^2} - \frac{2\rho\left(x-\mu_x\right)\left(y-\mu_y\right)}{\sigma_x\sigma_y}\right)\right]$$

and covariance structure $\Sigma = \begin{bmatrix} \sigma_x^2 & \rho\sigma_x\sigma_y \\ \rho\sigma_x\sigma_y & \sigma_y^2 \end{bmatrix}$

It is typical in the literature to predicate an SEM analysis with the caveat that one needs to make strong arguments for the complex models constructed from the unobserved, latent constructs tested with the particular SEM, in order to support the particular links that are included in the model. This is usually interpreted to mean that each proposed (and tested) link in the SEM needs to be supported with references to prior research, anecdotal evidence, and so forth. This may simply mean the wholesale import of a preexisting model (e.g., "theory or reasoned action" model or "technology acceptance model") based on the success of that model in other contexts, but not specifically building on the particular effects under investigation. But it is uncommon to see any discussion of the particular links (causal or otherwise) or combinations of links that are *excluded* (either implicitly or explicitly) from the SEM model. Ideally, there should also be similarly strong arguments made for the inapplicability of omitted links or omitted combinations of links.

We can formalize these observations by letting i be the number of the potential links between latent variables. Extend the individual link minimum sample size to a minimum sample size for the entire SEM, building up from pairs of latent variables by determining the number of possible combinations of the i pairs, each with an "effect" that needs detection. Each effect can be dichotomized:

$$\text{link}_i = \begin{cases} 0 : \rho_i < \delta \\ 1 : \rho_i \geq \delta \end{cases}$$

Our problem is to compute the number of distinct structural equation models that can exist in terms of the 0,1 values of their links using combinatorial analysis (Figs. 6.3 and 6.4).

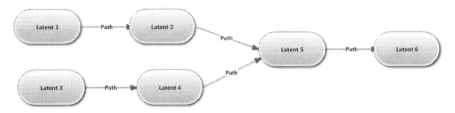

Fig. 6.3 An example of a structural equation model with six latent variables and five correlations

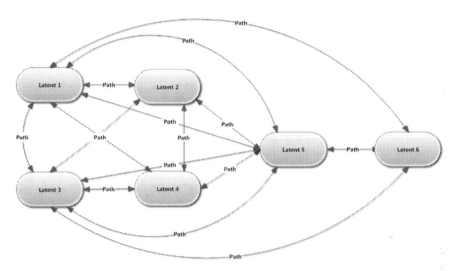

Fig. 6.4   The SEM example with all possible paired links shown

Then each combination of $\{0, 1\}$ values for links which our tests of the SEM on the whole require us to discriminate among provides us a set of $\dfrac{k(k-1)}{2}$ binary numbers, each representing a unique combination of latent variables. The unique model hypothesized in any particular study will be some model (binary number) which is exactly one out of the possible $2^{k(k-1)/2}$ ways of connecting these latent variables; testing must discriminate this path from the possible $2^{k(k-1)/2} - 1$ other paths which collectively define the alternative hypothesis.

For hypothesis testing with a significance of α^* (which we have by default set to $\alpha^* = 0.05$) on each link, it is necessary to correct for effective significance level α in differentiating one possible model from all other hypothesized structural equation models that are possible. The Šidàk correction is a commonly used alternative for the Bonferroni correction where an experimenter tests a set of hypotheses with a dataset controlling the family-wise error rate. In the context of the current research the Šidàk correction provides the most accurate results. For the following analysis, a Šidàk correction gives $\alpha = \alpha(k) = 1 - (1 - \alpha^*)^{2/k(k-1)}$ where the power of the test can be held at $1 - \beta = 0.8$ over the entire SEM with no modification.

6.9 Minimum Effect Size δ

Minimum effect, in the context of structural equation models, is the smallest correlation between latent variables that we wish to be able to detect with our sample and model. Small effects are more difficult to detect than large effects as they require more information to be collected. Information may be added to the analysis by collecting more sample observations, adding parameters, and constructing a better model (Fig. 6.5).

Sample size for hypothesis testing is typically determined from a *critical value* that defines the boundary between the rejection (set by α) and nonrejection (set by β) regions. The minimum sample size that can differentiate between H_0 and H_A occurs where the *critical value is exactly the same* under the null and alternative hypotheses. The approach to computing sample size here is analogous to standard univariate calculations (Cochran, 1977; Kish, 1995; Lohr, 1999; Snedecor & Cochran, 1989;

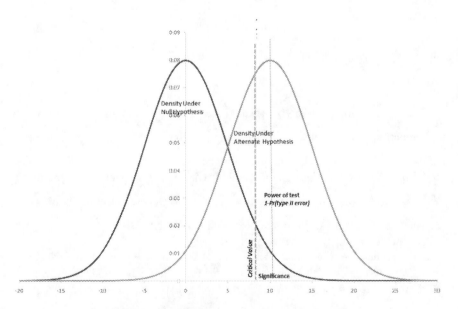

Fig. 6.5 Significance and power for the minimum effect that needs to be detected

Westland & See-to, 2007) but using a formulation for variance customized to this problem.

In the context of structural equation models, canonical correlation between latent variables should be seen simply as correlation, the "canonical" qualifier referring to the particulars of its calculation in SEM since the latent variables are unobserved, and thus cannot be directly measured. Correlation is interpreted as the strength of statistical relationship between two random variables obeying a joint probability distribution (Kendall & Gibbons, 1990) like a bivariate normal. Several methods exist to compute correlation: the Pearson's product moment correlation coefficient (Fisher, 1921, 1990) and Spearman's rho and Kendall's tau (Kendall & Gibbons, 1990) are perhaps the most widely used (Mari & Kotz, 2001). Besides these three classical correlation coefficients, various estimators based on M-estimation (Shevlyakov & Vilchevski, 2002) and order statistics (Schechtman & Yitzhaki, 1987) have been proposed in the literature. Strengths and weaknesses of various correlation coefficients must be considered in decision making. The Pearson coefficient, which utilizes all the information contained in the variates, is optimal when measuring the correlation between bivariate normal variables (Stuart & Ord, 1987). However, it can perform poorly when the data is attenuated by nonlinear transformations. The two rank correlation coefficients, Spearman's rho and Kendall's tau, are not as efficient as the Pearson correlation under the bivariate normal model; nevertheless they are invariant under increasing monotone transformations, thus often considered as robust alternatives to the Pearson coefficient when the data deviates from bivariate normal model. Despite their robustness and stability in non-normal cases, the M-estimator-based correlation coefficients suffer great losses (up to 63 % according to Xu, Hung, Niranjan, and Shen (2010)) of *asymptotic relative efficiency* to the Pearson coefficient for normal samples, though such heavy loss of efficiency might not be compensated by their robustness in practice. Schechtman and Yitzhaki (1987) proposed a correlation coefficient based on order statistics for the bivariate distribution which they call Gini correlation (because it is related to *Gini's mean difference* in a way that is similar to the relationship between Pearson correlation coefficient and the variance). As a measure of such strength, correlation should be large and positive if there is a high probability that large or small values of one variable occur (respectively) in conjunction with large of small values of another; and it should be large and negative if the direction is reversed (Gibbons & Chakraborti, 1992).

We will use a standard definition of minimum effect size to be detected—the strength of the relationship between two variables in a statistical population as measured by the correlation ρ for paired latent variables—following conventions articulated in Wilkinson and APA Task Force on Statistical Inference (1999); Nakagawa and Cuthill (2007), and Brand, Bradley, Best, and Stoica (2008). Where we are assessing completed research, we can substitute for δ the smallest correlation (effect size) on all of the links between latent variables in the SEM. Cohen (1988, 1992) provides the following guidelines for the social sciences: small effect size, $|\rho| = 0.1$–0.23; medium, $|\rho| = 0.24$–0.36; large, $|\rho| = 0.37$; or larger. Cohen's recommendations that $|\rho| = 0.37$ still leave room for a great deal of dispersion, and we might find it difficult to visually determine correlation merely by looking at a scatterplot where the variables on the two axes have correlation $|\rho| = 0.37$.

6.10   Estimator for Correlation in a Bivariate Normal Distribution

Let (Xi, Yi) $i = 1,2,\ldots,n$ be a random sample of independent and identically distributed (i.i.d.) data pairs of size n from the bivariate normal population of (X, Y) population with continuous joint cumulative distribution function. Let $X_{1:n}$ " $X_{2:n}$ " $\cdots$ " $X_{n:n}$ be the order statistics (where the first subscript is the rank, and the second the sample size) of the Xi sample values; let $Y_{1:n}$ " $Y_{2:n}$ " $\cdots$ " $Y_{n:n}$ be the order statistics of the Yi sample values; and let $Y_{[i:n]}$ be the Y sample value associated with the $X_{i:n}$ sample value in the sample pairs (Xi, Yi). $Y_{[i:n]}$ is called the concomitant of the ith order statistic (Balakrishnan & Rao, 1998). Reversing the roles of X and Y, we can also obtain the associated $X_{[i:n]}$. Extending the work of Schechtman and Yitzhaki (1987), Xu et al. (2010) show that the two Gini correlations with respect to (Xi, Yi) are

$$\hat{\rho}_{G_{XY}}(X,Y) = \frac{\dfrac{1}{n(n-1)}\sum_{i}^{n}(2i-n-1)X_{[i:n]}}{\dfrac{1}{n(n-1)}\sum_{i}^{n}(2i-n-1)X_{i:n}}$$

and

$$\hat{\rho}_{G_{YX}}(Y,X) = \frac{\dfrac{1}{n(n-1)}\sum_{i}^{n}(2i-n-1)Y_{[i:n]}}{\dfrac{1}{n(n-1)}\sum_{i}^{n}(2i-n-1)Y_{i:n}}$$

In general $\hat{\rho}_{G_{XY}}(X,Y)$ is not symmetric—that is, $\hat{\rho}_{G_{XY}}(X,Y) \neq \hat{\rho}_{G_{YX}}(Y,X)$. Such asymmetry violates the axioms of correlation measurement (Gibbons & Chakraborti, 1992; Mari & Kotz, 2001) which is assumed in SEM estimation. Xu et al. (2010) provide a symmetrical estimator (which we use here) obtained from their linear combination:

$$\hat{\rho}_{G}(Y,X) = \frac{1}{2}\left[\hat{\rho}_{G_{YX}}(Y,X) + \hat{\rho}_{G_{XY}}(X,Y)\right]$$

Gini correlation $\hat{\rho}_{G}$ possesses the following general properties (Schechtman & Yitzhaki, 1987):

1. $\hat{\rho}_G \in [-1,1]$.
2. $\hat{\rho}_G(Y,X) = \hat{\rho}_G(Y,X) = \pm 1$ if Y is a monotone increasing (decreasing) function of X.
3. $\hat{\rho}_G(Y,X)$ is asymptotically unbiased and the expectations of $\hat{\rho}_G(Y,X)$ and $\hat{\rho}_G(X,Y)$ are zero when Y is independent of X.
4. $\hat{\rho}_G(+,+) = -\hat{\rho}_G(-,+) = -\hat{\rho}_G(+,-) = \hat{\rho}_G(-,-)$ for both $\hat{\rho}_G(Y,X)$ and $= \hat{\rho}_G(X,Y)$.
5. $\hat{\rho}_G(Y,X)$ is invariant under all strictly monotone transformations of X.
6. $\hat{\rho}_G(Y,X)$ is scale and shift invariant with respect to both X and Y.
7. $\sqrt{n}(\hat{\rho}_G - \rho) \xrightarrow{\mathcal{D}} \mathbb{N}(0,\sigma_G^2)$, i.e., converges in distribution to a normal distribution with mean zero and variance σ_G^2 (this is from Schechtman and Yitzhaki (1987) applying methods developed by Hoeffding (1948)).
8. The Spearman rho measure of correlation is a special case of $\hat{\rho}_G(Y,X)$ (Xu et al., 2010).

Xu et al. (2010) showed that Gini correlations are asymptotically normal with the following mean and variance[1]:

$$\hat{\mu}_G = \mu(\hat{\rho}_G) =$$
$$\rho - \frac{2(n-2)}{n(n-1)}\left(\arcsin\left(\frac{\rho}{2}\right) + \rho\sqrt{4-\rho^2} - \rho\sqrt{3}\right) +$$
$$\frac{\pi}{3}\frac{\rho(n+1)}{n(n-1)} + \frac{2}{n(n-1)}\left(\arcsin(\rho) + \rho\sqrt{1-\rho^2}\right) + o(n^{-1})$$

$$\hat{\sigma}_G^2 = \sigma^2(\hat{\rho}_G) =$$
$$\frac{(1-\rho^2)}{n(n-1)}\left(\sqrt{1-\rho^2} - \rho\arcsin(\rho) + \frac{\pi(n+1)}{6}\right) +$$
$$\frac{(n-2)(1-\rho^2)}{n(n-1)}\left(\frac{(1-\rho^2)}{\sqrt{4-\rho^2}} - \rho\arcsin\left(\frac{\rho}{2}\right)\right) + o(n^{-1})$$

Xu et al. (2010) used Monte Carlo simulations to verify these formulas' asymptotic results (using *asymptotic relative efficiency* and *root mean square error* performance metrics) showing that they are applicable for data of even relatively small sample sizes (down to around 30 sample points). Their simulations confirmed and extended Hea and Nagarajab's (2009) Monte Carlo simulations exploring the behavior of nine distinct correlation estimators of the bivariate normal correlation coefficient, including the estimator $\hat{\rho}_G$, sample correlation for the bivariate normal, and estimators based on order statistics. The estimator $\hat{\rho}_G$ was found generally to

---

[1] $o(n^{-1})$ convergence implies that for the remaining terms $v(n)$ go to zero faster than n^{-1}; $nv(n) \xrightarrow[n\to\infty]{} 0$.

reduce bias and improve efficiency as well or be better than other correlation estimators in the study. Xu et al. (2010) also compared $\hat{\rho}_G$ with three other closely
related correlation coefficients: (1) classical Pearson's product moment correlation
coefficient, (2) Spearman's rho, and (3) order statistics correlation coefficients. Gini
correlation bridges the gap between the order statistics correlation coefficient and
Spearman's rho, and its estimators are more mathematically tractable than
Spearman's rho, whose variance involves complex elliptic integrals that cannot be
expressed in elementary functions. Their efficiency analysis showed that estimator
$\hat{\rho}_G$'s loss of efficiency is between 4.5 and 11.3 %, much less than that of Spearman's
rho which ranges from 8.8 to 30.5 %.

6.11   Calculation of Sample Size on a Single Link

Construct a hypothesis test to just detect the minimum effect size δ:

$$H_0 : \rho - \rho_0 = 0$$

$$H_A : \rho - \rho_0 = \delta$$

The one-sample, two-sided formulation that reconciles the null and alternative
hypothesis tests for the estimator $\hat{\rho}_G \equiv \hat{\rho}_G(n)$ is

$$0 + z_{1-\alpha/2} \hat{\sigma}_G(n) = \delta + z_{1-\beta} \hat{\sigma}_G(n)$$

Xu et al. (2010) show that $|\hat{\mu}_G - \rho| \xrightarrow[n \to \infty]{} 0$ quickly: for $n > 30$ from a bivariate
normal population they show that we can assume $|\hat{\mu}_G - \rho| = 0$. Similarly, for $n > 30$
we can assume that z-values are adequate approximations for t-values in the formula. Even under the very weak assumptions of the "rule of 10" a sample of $n = 30$
implies a model of at most three variables—significantly simpler than the majority
of published models. Rearranging to place all terms with n on the left-hand side

$$\hat{\sigma}_G^2(n) = \left(\frac{\delta}{z_{1-\alpha/2} - z_{1-\beta}} \right)^2 \equiv H$$

Thus to within little $o(n^{-1})$ and using the formula for $\hat{\sigma}_G^2$

$$H \cong f(n,\rho) = \frac{(1-\rho^2)}{n(n-1)} \left(\sqrt{1-\rho^2} - \rho \arcsin(\rho) + \frac{\pi(n+1)}{6} \right.$$
$$\left. + \frac{(n-2)(1-\rho^2)}{n(n-1)} \left(\frac{(1-\rho^2)}{\sqrt{4-\rho^2}} - \rho \arcsin\left(\rho/2\right) \right) \right)$$

We want to restate this as some function that calculates sample size $n = g(H, \rho)$. Solve for n by simplifying in terms of

$$A = 1 - \rho^2$$

$$B = \rho \arcsin\left(\rho/2\right)$$

$$C = \rho \arcsin(\rho)$$

$$D = \frac{A}{\sqrt{3 - A}}$$

$$H = \left(\frac{\delta}{z_{1-\alpha/2} - z_{1-\beta}}\right)^2$$

Then $n = \dfrac{-E \pm \sqrt{E^2 - 4F}}{2}$ are the solutions for the quadratic equation that restates $H - f(n, \rho) \overset{\bullet}{=} 0$:

$$n^2 - \frac{A\left(\frac{\pi}{6} - B + D\right) + H}{H} n - \frac{A\left(\frac{\pi}{6} + \sqrt{A} + 2B - C - 2D\right)}{H} = n^2 + En + F = 0$$

Or in terms of A, B, C, D, and H and taking the largest root

$$n = \frac{1}{2H}\left(A\left(\frac{\pi}{6} - B + D\right) + H + \sqrt{\left[A\left(\frac{\pi}{6} - B + D\right) + H\right]^2 + 4AH\left(\frac{\pi}{6} + \sqrt{A} + 2B - C - 2D\right)}\right)$$

This then provides us with *necessary* conditions for sample adequacy in SEM-based hypothesis testing. Combining with Boomsma's criterion, we can assert that there are at least two necessary conditions for sample adequacy:

1. The sample size needed to compensate for the ratio of number of indicator variables to latent variables (summarized from Monte Carlo simulations that have appeared in the literature).
2. The sample size required to assure the existence or nonexistence of a minimum effect (correlation) on each possible pair of latent variables in the SEM (determined analytically).

Of course, neither of these conditions is *sufficient* to assure sample adequacy for a particular choice of (α, β) because there are so many other factors that can affect estimation and sample size—multicollinearity, appropriateness of datasets, and so forth. Additionally, the information contained in the sample and indicator variables must be adequate to compensate for variations in particular SEM estimation methodologies. For example, PLS-PA approaches generate parameter estimates that lack consistency. Dhrymes (1970); Schneeweiß (1990, 1991, 1993); Thomas, Lu, and Cedzynski (2005); and Fèhèr (1989) all demonstrate that the IV/2SLS techniques converge to the same estimators, but are more robust. Jöreskog (Jöreskog, 1967, 1970; Jöreskog & Sörbom, 1996) suggests that departures from normal distribution for the indicators will demand larger samples, and that non-normal indicators require one, two, or three magnitudes larger samples, depending on distribution (Fig. 6.6).

These are *absolute* minimum sample sizes—applicable exactly only when very specific conditions exist, such as normalcy of data (which is not the case for Likert data). It should be noted that these bounds are calculated assuming that data is normally distributed; for non-normal data, sample sizes one to two magnitudes larger may be needed. A minimum effect larger than the significance level would likely be suspected; thus researchers would be likely to choose sample sizes to the left of 0.05 on the x-axis. SEM algorithms often scale the coefficients, where scale may be set through the values chosen for factor weights.

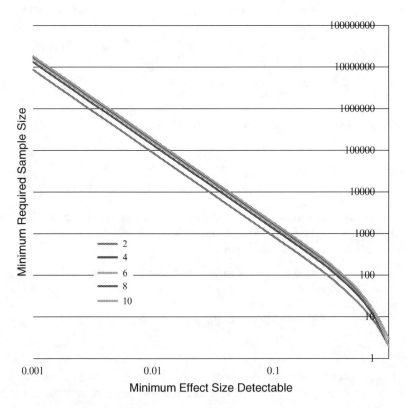

Fig. 6.6 Lower bound on sample size in (Westland, 2010) 0.05 significance. 0.8 power and minimum detectable effect

The network structure of SEM adds an additional complication to computing minimum detectable effect. That effect must be detected across all possible links, not just the ones that are assumed in the research model. Soper (2012) provides software for this sample size calculation. These provide a distinct set of necessary conditions for sample adequacy to Boomsma's results. In some cases, Boomsma's simulations will constrain the lower bound on the sample size; in others, these constrain the lower bound.

6.12   Can Resampling Recover Information Lost Through Likert Mapping?

The loss of information in the Likert mapping from a continuous set of beliefs into a very simple, discrete Likert categorical distribution is likely to be both substantial and difficult to measure. Subjects may not have strongly held beliefs, and where they do, these beliefs can change quickly under the influence of new data (e.g., consider clients' perceptions about Arthur Andersen after their failures at Enron became public).

Resampling or "bootstrapping" is one method—but not one without controversy—for attempting to gather more information about the actual perceptions being approximated in the Likert scale. The statistical use of "bootstrapping" was borrowed from *The Surprising Adventures of Baron Munchausen* (Raspe, 2004), where Baron Munchausen pulls himself out of a swamp by his bootstraps. Bootstrapping algorithms are built into most SEM software, and are responsible for many of their desirable small sample properties. Resampled data provides a simple, straightforward way to derive estimates of standard errors and confidence intervals for complex estimators of complex parameters of the distribution, such as are typical of SEM. Like Monte Carlo approaches, it is an appropriate way to control and check the stability of the results.

When applied properly, it is a useful tool for developing and incorporating model assumptions that are consistent with the data. Bootstrap data is asymptotically consistent, but does not provide general finite-sample assurances. Moreover, there is no guarantee that bootstrapped Likert data will better represent underlying beliefs than the original data. Bootstrapping does not create new information; if the researcher is lucky, it may provide modeling insights that were not previously available—somewhat like a pretest.

The basic idea behind bootstrapping is that the sample we have collected is often the best guess we have as to the shape of the population from which the sample was taken. Thus we could assume (without actually collecting data) that future data will come from this empirical distribution, and artificially generate more data. We can use similar ideas for imputation of missing data, and all of this falls under the broader rubric of resampling. Rather than making assumptions directly—for example, that the data is drawn from a normal population—we can let the bootstrap-generated data introduce this assumption into the modeling. It estimates the sampling distribution of an estimator by sampling with replacement from the original sample, most often with the purpose of deriving robust estimates of standard errors and confidence intervals of a population parameter like a mean, median, and proportion. Jackknifing is similar to bootstrapping, and is used in statistical inference to estimate the bias and standard error of a statistic, when a random sample of observations is used to calculate it. The basic idea behind the jackknife variance estimator lies in systematically recomputing the statistic estimate leaving out one or more observations at a time from the sample set.

6.13   Data Screening

Prior to the descriptive and positive testing, it is important to screen the data. This is the first step towards formalization of the research. The order of the screening is important as decisions at the earlier steps influence decisions to be taken at later steps. For example, if the data is both non-normal and has outliers, the decision to delete values or transform the data is confronted. Transformation of the variable is usually preferred as it typically reduces the number of outliers, and is more likely to produce normality, linearity, and homoscedasticity; in social science work, data is often limited to positive values only, and may be ordinal as well, as is the case for Likert scale responses. Screening will aid in the isolation of data peculiarities and allow the data to be adjusted in advance of further multivariate analysis. Tabachnick and Fidell (Merton, 1988; Tabachnick & Fidell, 1989) suggest the following data screening tasks:

1. Inspect univariate descriptive statistics for accuracy of input:

    (a) Out-of-range values, be aware of measurement scales
    (b) Plausible means and standard deviations
    (c) Coefficient of variation

2. Evaluate the amount and distribution of missing data: deal with problem.
3. Independence of variables.
4. Identify and deal with non-normal variables.

    (a) Check skewness, kurtosis, and probability plots.
    (b) Transform variables (if desirable).
    (c) Check results of transformations.

5. Identify and deal with outliers:

    (a) Univariate outliers
    (b) Multivariate outliers

6. Check pairwise plots for nonlinearity and heteroscedasticity.

7. Evaluate variables for multicollinearity and singularity.
8. Check for spatial autocorrelation.

Fortunately, there are excellent tools available to the researcher for data screening. PRELIS (a part of the LISREL statistical package that can also be made to work with PLS packages) and its SPSS counterparts (which are often used by themselves or for preliminary testing, transformation, and culling of data in PLS) in AMOS provide automated support for all of these screening tasks. The *R-language* statistical package supports many data screening tasks that allows for "eyeballing" data for patterns, and transforming, plotting residuals, and other useful functions on the fly. Without proper data screening, data exploration and testing later in the research cycle cannot be relied upon to provide reliable answers to research questions.

6.14   Exploratory Specification Search

Leamer (1978) surveyed statistical *specification search* approaches (in his case, Bayesian methods) that can be used for preliminary model building, model re-specification and ad hoc inference with weak and/or nonexperimental data. Such specification searches are the main task of hypothesis justification, pretests, and suggestions for extending already completed research. They play an especially important role in the social sciences where core constructs are often not directly observable (e.g., consumer satisfaction) and where two important classes of data— survey and economic—are nonexperimental. Iteratively combining (1) factor analysis and other cluster analysis methods with (2) PLS path analysis for holistically exploring the causal relationships between "clusters" (the first step towards generating theory-driven latent factor constructs), and (3) stepwise regression for decisions on whether or not to include factors, provides comprehensive exploration of the parameter space supported by a particular set of data and prior beliefs. Clearly the results cannot be seen as ends in themselves. Rather this exploratory specification search assures—in moving to the rigorous confirmatory testing of a particular model—that any feasible models, variables, and causal relationships that should be tested are included.

When SEM are built around valid real-world constructs (even if these are unobservable) the algorithms proposed in this chapter impose only weak additional assumptions on the indicators and latent variables in order to compute sample sizes adequate for estimation. Our limited application to a window of IS and e-commerce publications has shown that concerns are warranted concerning existing SEM sample size calculations and we need to remain suspicious of conclusions reached in studies based on inadequate sample sizes. Furthermore, a large number of studies in our sample devised their tests without first committing to minimum effect size that they were trying to detect, or indicated in portion of nonresponse in surveys. It is clear that journal referees need to begin asking for survey response, minimum effect size δ, and a justification of the sample size. By incorporating these suggestions, it is argued that the research community will enhance the credibility and applicability of their research, with a commensurate improved impact and influence in both industry and academe.

References

Balakrishnan, N., & Rao, C. R. (1998). *Order statistics: Applications* (Handbook of statistics, Vol. 17). New York, NY: Elsevier.

Barclay, D. W., Higgins, C., & Thompson, R. (1995). The partial least squares (PLS) approach to causal modeling: Personal computer adaptation and use as an illustration. *Technology Studies, 2*(2), 285–309.

Bentler, P. M. (1989). *EQS, structural equations, program manual, program version 3.0* (p. 6). Los Angeles, CA: BMDP Statistical Software.

Bollen, K. A. (1989). *Structural equations with latent variables* (p. 268). New York, NY: Wiley.

Boomsma, A. (1982a). Robustness of LISREL against small sample sizes in factor analysis models. In K. G. Joreskog & H. Wold (Eds.), *Systems under indirect observations, causality, structure, prediction (Part 1)* (pp. 149–173). Amsterdam, The Netherlands: North Holland.

Boomsma, A. (1982b). The robustness of LISREL against small sample sizes in factor analysis models. In K. G. Jöreskog & H. Wold (Eds.), *Systems under indirect observation: Causality, structure, prediction* (Vol. 149, pp. 149–173). Amsterdam, The Netherlands: North-Holland.

Brand, A., Bradley, M. T., Best, L. A., & Stoica, G. (2008). Accuracy of effect size estimates from published psychological research. *Perceptual and Motor Skills, 106*(2), 645–649.

Browne, M. W., & Cudeck, R. (1989). Single sample cross-validation indices for covariance structures. *Multivariate Behavioral Research, 24*, 445–455.

Browne, M. W., & Cudeck, R. (1993). Alternative ways of assessing model fit. In K. A. Bollen & J. S. Long (Eds.), *Testing structural equation models* (pp. 136–162). Newbury Park, CA: Sage.

Cattell, R. B. (1966). *Handbook of multivariate experimental psychology*. Chicago, IL: Rand McNally.

Chin, W. W. (1998). The partial least squares approach to structural equation modeling. In G. A. Marcoulides (Ed.), *Modern methods for business research* (pp. 295–336). Mahwah, NJ: Lawrence Erlbaum Associates.

Chin, W. W., & Newsted, P. R. (1999). Structural equation modeling analysis with small samples using partial least squares. In R. Hoyle (Ed.), *Statistical strategies for small sample research* (pp. 307–341). Thousand Oaks, CA: Sage.

Cliff, N. (1988). The eigenvalues-greater-than-one rule and the reliability of components. *Psychological Bulletin, 103*(2), 276.

Cochran, W. G. (1977). *Sampling techniques* (3rd ed.). New York, NY: Wiley.

Cohen, J. (1988). *Statistical power analysis for the behavioral sciences* (2nd ed.). Hillsdale, NJ: Lawrence Erlbaum Associates.

Cohen, J. (1992). A power primer. *Psychological Bulletin, 112*, 155–159.

Copas, J. B., & Li, H. G. (1997). Inference for non-random samples. *Journal of the Royal Statistical Society, Series B (Statistical Methodology), 59*(1), 55–95.

Cronbach, L. J., & Meehl, P. E. (1955). Construct validity in psychological tests. *Psychological Bulletin, 52*(4), 281.

Dhrymes, P. J. (1970). *Econometrics—Statistical foundations and applications* (p. 53). New York, NY: Evanston.

Dijkstra, T. (1983). Some comments on maximum likelihood and partial least squares methods. *Journal of Econometrics, 22*(1), 67–90.

Ding, L., Belicer, W. F., & Harlow, L. L. (1995). The effects of estimation methods, number of indicators per factor and improper solutions on structural equation modeling fit indices. *Structural Equation Modeling, 2*, 119–144.

Fava, J. L., & Velicer, W. F. (1992a). An empirical comparison of factor, image, component, and scale scores. *Multivariate Behavioral Research, 27*(3), 301–322.

Fava, J. L., & Velicer, W. F. (1992b). The effects of overextraction on factor and component analysis. *Multivariate Behavioral Research, 27*(3), 387–415.

Fèhèr, K. (1989). *Comparison of LISREL and PLS estimation methods in latent variable models. Introducing latent variables into econometric models* (Manuscript SFB, Vol. 303). Bonn, Germany: University of Bonn.

Fisher, R. A. (1921). On the 'probable error' of a coefficient of correlation deduced from a small sample. *Metron, 1,* 3–32.

Fisher, R. A. (1935). *The design of experiments.* Edinburgh, UK: Oliver & Boyd.

Fisher, R. A. (1990). *Statistical methods, experimental design, and scientific inference.* New York, NY: Oxford University Press.

Ford, E. D. (2000). *Scientific method for ecological research.* Cambridge, UK: Cambridge University Press.

Friedman, M. (1953). *Essays in positive economics.* Chicago, IL: University of Chicago Press.

Gerbing, D. W., & Anderson, J. C. (1985). The effects of sampling error and model characteristics on parameter estimation for maximum likelihood confirmatory factor analysis. *Multivariate Behavioral Research, 20,* 255–271.

Geweke, J. F., & Singleton, K. J. (1980). Interpreting the likelihood ratio statistic in factor models when sample size is small. *Journal of the American Statistical Association, 7*(369), 133–137.

Gibbons, J. D., & Chakraborti, S. (1992). *Nonparametric statistical inference* (3rd ed.). New York, NY: Marcel Dekker.

Goodhue, D., Lewis, W., & Thompson, R. (2006). *PLS, small sample size, and statistical power in MIS Research,* HICSS (Vol. 8, p. 202b). Proceedings of the 39th Annual Hawaii International Conference on System Sciences (HICSS'06).

Goodhue, D., William, L., & Thompson, R. (2007). Statistical power in analyzing interaction effects: Questioning the advantage of PLS with product indicators (research note). *Information Systems Research, 18*(2), 211–227.

Hea, Q., & Nagarajab, H. N. (2009). Correlation estimation using concomitants of order statistics from bivariate normal samples. *Communications in Statistics - Theory and Methods, 38*(12), 2003–2015.

Hoeffding, W. (1948). A class of statistics with asymptotically normal distribution. *Annals of Mathematical Statistics, 19,* 293–325.

Jöreskog, K. G. (1967). Some contributions to maximum likelihood factor analysis. *Psychometrika, 32*(4), 443–482.

Jöreskog, K. G. (1970). A general method for analysis of covariance structures. *Biometrika, 57,* 239–251.

Jöreskog, K. G., & Sörbom, D. (1996). *LISREL 8 user's reference guide.* Chicago, IL: Scientific Software International.

Joreskog, K. G., Sorbom, D., & Magidson, J. (1979). *Advances in factor analysis and structural equation models.* Cambridge, MA: Abt Books.

Kahai, S. S., & Cooper, R. B. (2003). Exploring the core concepts of media richness theory: The impact of cue multiplicity and feedback immediacy on decision quality. *Journal of Management Information Systems, 20*(1), 263–299.

Kaiser, H. F. (1960). The application of electronic computers to factor analysis. *Educational and Psychological Measurement, 20,* 141–151.

Kaiser, F. H. (1991). Coefficient alpha for a principal component and the Kaiser-Guttman rule. *Psychological Reports, 68*(3), 855–858.

Kaiser, H. F. (1992). On Cliff's formula, the Kaiser-Guttman rule, and the number of factors. *Perceptual and Motor Skills, 74*(2), 595–598.

Kendall, M., & Gibbons, J. D. (1990). *Rank correlation methods* (5th ed.). New York, NY: Oxford University Press.

Kish, L. (1995). *Survey sampling.* New York, NY: Wiley.

Koopmans, T. C. (1951). Analysis of production as an efficient combination of activities. *Activity Analysis of Production and Allocation, 13*, 33–37.

Koopmans, T. C. (1957). *Three essays on the state of economic science* (Vol. 21). New York, NY: McGraw-Hill.

Koopmans, T. C. (1963). *Appendix to 'On the concept of optimal economic growth'*. New Haven, CT: Cowles Foundation for Research in Economics, Yale University.

Lauro, C., & Vinzi, V. E. (2002). *Some contributions to PLS path modeling and a system for the European customer satisfaction* (Atti della XL1 riunione scientifica SIS). Milano, Italy: Universita di Milano Bicocca.

Leamer, E. E. (1978). *Specification searches*. New York, NY: Wiley.

Likert, R. (1932). A technique for the measurement of attitudes. *Archives of Psychology, 22*, 55.

Likert, R., Roslow, S., & Murphy, G. (1934). A simple and reliable method of scoring the Thurstone attitude scales. *Journal of Social Psychology, 5*(2), 228–238.

Lohr, S. L. (1999). *Sampling: Design and analysis*. Pacific Grove, CA: Duxbury.

Mari, D. D., & Kotz, S. (2001). *Correlation and dependence*. London: Imperial College Press.

Marsh, H. W., & Bailey, M. (1991). Confirmatory factor analyses of multitrait-multimethod data: A comparison of alternative models. *Applied Psychological Measurement, 15*(1), 47–70.

Marsh, H. W., Balla, J. R., & Hau, K.-T. (1996). An evaluation of incremental fit indices: A clarification of mathematical and empirical properties. In G. A. Marcoulides & R. E. Schumacker (Eds.), *Advanced structural equation modeling: Issues and techniques* (pp. 315–353). Mahwah, NJ: Lawrence Erlbaum Associates, Inc.

Marsh, H. W., Balla, J. R., & McDonald, R. P. (1988). Goodness of fit indexes in confirmatory factor analysis: The effect of sample size. *Psychological Bulletin, 103*, 391–410.

Marsh, H. W., Hau, K.-T., Balla, J. R., & Grayson, D. (1998). Is more ever too much? The number of indicators per factor in confirmatory factor analysis. *Multivariate Behavioral Research, 33*, 181–220.

Merton, R. K. (1988). The Matthew effect in science, II: Cumulative advantage and the symbolism of intellectual property. *Isis, 79*, 606–623.

Nakagawa, S., & Cuthill, I. C. (2007). Effect size, confidence interval and statistical significance: A practical guide for biologists. *Biological Reviews of the Cambridge Philosophical Society, 82*, 591–605.

Nunnally, J. C. (1967). *Psychometric theory* (p. 355). New York, NY: McGraw-Hill.

Podsakoff, P. M., & Organ, D. W. (1986). Self-reports in organization research: Problems and prospects. *Journal of Management, 12*, 531–544.

Raspe, R. E. (2004). *The surprising adventures of Baron Munchausen*. Whitefish, MT: Kessinger Pub.

Schechtman, E., & Yitzhaki, S. (1987). A measure of association based on Gini's mean difference. *Communications in Statistics - Theory and Methods, 16*, 207–231.

Schneeweiß, H. (1990). Models with latent variables: LISREL versus PLS. *Contemporary Mathematics, 112*, 33–40.

Schneeweiß, H. (1991). Models with latent variables: LISREL versus PLS. *Statistica Neerlandica, 45*, 145–157.

Schneeweiß, H. (1993). Consistency at large in models with latent variables. In K. Hagen, D. J. Barthdomew, & M. Deistler (Eds.), *Statistical modelling and latent variables* (pp. 299–320). Amsterdam, The Netherlands: Elsevier.

Shevlyakov, G. L., & Vilchevski, N. O. (2002). *Robustness in data analysis: Criteria and methods* (Modern probability and statistics). Utrecht, The Netherlands: VSP.

Snedecor, G. W., & Cochran, W. G. (1989). *Statistical methods* (8th ed.). Ames, IA: Iowa University Press.

Soper, D. (2012). A priori sample size calculator for structural equation models. http://www.danielsoper.com/statcalc3/calc.aspx?id=89

Stuart, A., & Keith Ord, J. (1987). *Kendall's advanced theory of statistics*. New York, NY: Oxford University Press.

Tabachnick, B. G., & Fidell, L. S. (1989). *Using multivariate statistics*. New York, NY: Harper and Row.

Tanaka, J. S. (1987). "How big is big enough?": Sample size and goodness of fit in structural equation models with latent variables. *Child Development, 58*, 134–146.

Thomas, D. R., Lu, I. R. R., & Cedzynski, M. (2005). *Partial least squares: A critical review and a potential alternative*. Proceedings of the Annual Conference of Administrative Sciences Association of Canada, Management Science Division, Toronto.

Velicer, W. F., & Fava, J. L. (1998). Affects of variable and subject sampling on factor pattern recovery. *Psychological Methods, 3*(2), 231.

Westland, J. C. (2010). Lower bounds on sample size in structural equation modeling. *Electronic Commerce Research and Applications, 9*(6), 476–487.

Westland, J. C., & See-To, W. K. (2007). The short-run price-performance dynamics of microcomputer technologies. *Research Policy, 36*(5), 591–604.

Wilkinson, L., & APA Task Force on Statistical Inference. (1999). Statistical methods in psychology journals: Guidelines and explanations. *American Psychologist, 54*, 594–604.

Wolfram, S. (2002). *A new kind of science* (Vol. 5). Champaign, IL: Wolfram media.

Xu, W., Hung, Y. S., Niranjan, M., & Shen, M. (2010). Asymptotic mean and variance of Gini correlation for bivariate normal samples. *IEEE Transactions on Signal Processing, 58*(2), 522–534.

Chapter 7
Survey and Questionnaire Data

Surveys study various characteristics of individuals from a population. Modern surveys grew out of census procedures dating back to the Romans. Today, they are more often directed towards assessing the sentiment of a large population through marketing research, public opinion polls, epidemiological surveys, and various national economic, tax, and consumption surveys. Surveys are an essential part of managing complex bureaucracies of business, government, and public health.

Surveys ask questions to assess constructs such as preferences (e.g., for a tax cut), opinions (e.g., whether drugs are harmful), behavior (e.g., whether trust encourages purchasing), or facts (e.g., family size). The success of a survey depends on the representativeness of the sample with respect to a population of interest to the research question. Concerns in surveys range from questionnaire design, execution and interpretation (including follow-up on nonresponse, and adjustment for sample bias), sample design, data collection instruments, statistical adjustment of data, and data processing, systematic, and random survey errors. Cost constraints are imposed by weighing the cost of a data item against the cost of survey error and quality.

Questionnaires are an exceptionally popular survey instrument, and one unfortunately that tends to be too often collected and analyzed carelessly, laxly, and incorrectly. Questionnaires have advantages over other survey approaches in that they are cheap and require little effort to implement compared to, for example, verbal or telephone surveys, or interrogative protocols. Questionnaires suffer from many of the same shortcomings as other types of opinion polls.

Questionnaires designed for personal interrogation, such as the widely used Myers-Briggs Type Indicator, strictly limit responses, where respondents can answer either option but must choose only one response. Myers and Briggs, a home-schooled mother and daughter team that popularized typological theories proposed by Carl Gustav Jung in 1921, constructed a relatively ad hoc instrument that remains a bit fuzzy about what question it is actually intended to answer. Such ambiguity is not unusual in questionnaire design. Additionally, most questionnaires suffer significant

© Springer International Publishing Switzerland 2015
J.C. Westland, *Structural Equation Models*, Studies in Systems,
Decision and Control 22, DOI 10.1007/978-3-319-16507-3_7

nonresponse, especially with mail and online questionnaires; alternatively, people may not care enough to respond truthfully.

Interrogation techniques evolved significantly with the widespread adoption of the polygraph in police and government agencies after the 1930s. Polygraphs, or lie detectors, measure and record physiological indices such as blood pressure, pulse, respiration, and skin conductivity while the subject is asked and answers a series of questions. The theory is that with proper questionnaire design, dishonest answers will produce physiological responses different from truthful answers.

Polygraph testing typically begins with a pretest interview to gain some preliminary information which will later be used to develop diagnostic questions. This will be followed by a request for respondents to deliberately lie in order to calibrate the "dishonest" physiological parameters. Then the actual test starts. Some of the questions asked are irrelevant, others are diagnostic questions, and the remainder are relevant questions. The goal is not just to get responses, but to get true responses, and be able to identify casual or untrue responses.

Though polygraph protocols have much to inform survey questions applied elsewhere, these are often ignored because they are considered too much work by many social researchers. Instead, modern marketing, social science, and other survey research disciplines like to provide questionnaires where all questions are considered directly relevant, and where all responses are immediately quantified on a Likert scale, so that they can be fed into computer software to generate tables of statistics. The remainder of this chapter investigates the consequences of such an insouciant approach to survey research.

7.1 Rensis Likert's Contribution to Social Research

Likert scales are named for Rensis Likert[1] who developed them in his PhD thesis, and promoted the 1–5 Likert scale for the remainder of his career. Likert was a founder of the University of Michigan's Institute for Social Research, was the director from its inception in 1946 until 1970, and later founded a consulting firm to promote the Likert scale.

The purpose of a Likert (rating) scale is to allow respondents to express both the direction and strength of their opinion about a topic. A Likert item is simply a statement which the respondent is asked to evaluate according to any kind of subjective or objective criteria; generally the level of agreement or disagreement is measured. It is considered symmetric or "balanced" because there are equal amounts of positive and negative positions.

Often, researchers would prefer respondents to make a definite choice rather than choose neutral or intermediate positions on a scale. For this reason, a scale without a midpoint would be preferable, provided it does not affect the validity or reliability

---

[1] Despite his influence, Likert's name is often mispronounced with a long "i" ("lie-kurt")—Likert himself pronounced his name "lick-urt" with a short "i."

of the responses. Numerous studies have demonstrated that as the number of scale steps is increased, respondents' use of the midpoint category decreases (J. Friedman, Hastie, Rosset, Tibshirani, & Zhu, 2004; J. Friedman, Hastie, & Tibshirani, 2010; Komorita, 1963; Komorita & Graham, 1965; Matell & Jacoby, 1972; Sterne, Smith, & Cox, 2001; Wildt & Mazis, 1978).

Designing a scale with balanced keying (an equal number of positive and negative statements) can obviate the problem of acquiescence bias, since acquiescence on positively keyed items will balance acquiescence on negatively keyed items, but there are no widely accepted solutions to central tendency and social desirability biases.

When a Likert scale approximates an interval-level measurement, we can summarize the central tendency of responses using either the median or the mode, with "spread" measured by standard deviations, quartiles, or percentiles. Characteristics of the sample can be obtained from nonparametric tests such as chi-squared test, Mann–Whitney test, Wilcoxon signed-rank test, or Kruskal–Wallis test (Jamieson, 2004).

Likert items are considered symmetric or "balanced" where there are equal amounts of positive and negative positions. Rensis Likert used five ordered response levels, but seven and even nine levels are common as well. Allen and Seaman (2007) concluded that a 5- or 7-point scale may produce slightly higher mean scores relative to the highest possible attainable score, compared to those produced from a 10-point scale, and this difference was statistically significant. In terms of the other data characteristics, there was very little difference among the scale formats in terms of variation about the mean, skewness, or kurtosis.

7.2   Likert Scales

Likert scales are the most widely used approach to scaling responses in survey research, such that the term is often used interchangeably with rating scale. There is considerable discussion as to the exact meaning of Likert scaling, so much so that this is beyond the scope of this book. The Rasch model is the most intuitive, but not every set of Likert-scaled items can be used for Rasch measurement. The data has to be thoroughly checked to fulfill the strict formal axioms of the model (T. G. Bond & Fox, 2007). Likert scale data can, in principle, be used as a basis for obtaining interval-level estimates on a continuum by applying the polytomous Rasch model, when data can be obtained that fit this model. In addition, the polytomous Rasch model permits testing of the hypothesis that the statements reflect increasing levels of an attitude or trait, as intended. For example, application of the model often indicates that the neutral category does not represent a level of attitude or trait between the disagree and agree categories (Fig. 7.1).

Alternatively, a Likert scale can be considered as a grouped form of a continuous scale. This is important in SEM, since you implicitly treat the variable as if it were continuous for correlational analysis. Likert scales are clearly ordered category

Fig. 7.1 The Likert scoring
process (value × degree of
belief); continuously varying
strength of belief is
approximated with 5 levels of
belief

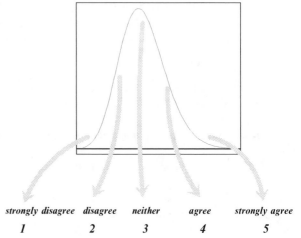

strongly disagree	*disagree*	*neither*	*agree*	*strongly agree*
1	*2*	*3*	*4*	*5*

scales, as required for correlational work, and the debate among methodologists is
whether they can be treated as equal interval scales (Fig. 7.2).

There are two main questions relating to the SEM use of Likert scale datasets.
The first one seeks to know the nature of Likert scale and if they can be used for correlation and chi-square test. Unfortunately, correlations from two belief distributions—
for example (1) belief about whether one is healthy and (2) belief about whether
one's career is successful—differ from the correlation of their Likert scale representations. Information is lost in the mapping to a Likert scale; how much information
is lost is probably unknowable in most cases (though I address this question from a
purely mechanistic standpoint later in the chapter). There is a body of research
(Kühberger, 1995) that concludes that people do not generally hold strong, stable,
and rational beliefs, and that their responses are very much influenced by the way in
which decisions are framed (which should serve as a strong caveat for the meticu-
lous design of survey instruments).

The second question relates to resolution or granularity of measurement.
Measurement in research consists in assigning numbers to entities otherwise called
concepts in compliance with a set of rules. These concepts may be "physical," "psychological," and "social." The concept *length* is physical. But the question remains,
"if I report length as 6 ft in a case, what exactly does that mean?" Even with physical
scales, there is an implied granularity; if I say that something is 6 ft long, this implies
less precision than length of 183 cm. In scientific pursuits, finer granularities can be
pursued to almost unimaginable levels—for example, the international standard for
length, adopted in 1960, is derived from the *2p10-5d5* radiation wavelength of the
noble gas *krypton-86*. The influence of choice of measuring stick on the results of
modeling is responsible for phenomena such as Benford's law (Gurevich, 1961) and
fractal scaling (Basmann, 1963).

With a choice of a Fisher information metric, we can explore the implications
of the widespread assumption that survey subject beliefs or other phenomena are

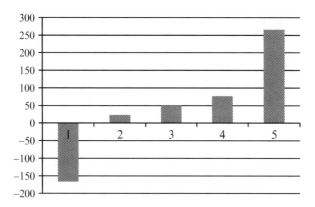

Fig. 7.2 Mapping of a belief distribution with [−200, +300] support to a 5-point Likert scale

normally distributed, but where they are measured and analyzed as Likert-scaled data. Likert scales allow respondents to express both the direction and strength of their opinion about a topic (Likert, 1974; Likert, Roslow, & Murphy, 1934). Thus a Likert item is a statement that the respondent is asked to evaluate according to any kind of subjective or objective criteria; generally the level of agreement or disagreement is measured. Survey researchers often impose various regularity conditions on the metrics implied in the construction of their survey instruments to eliminate biases in observations, and help assure that there is a proper matching of survey results and the analysis (Clarke, Worcester, Dunlap, Murray, & Bradley-Klug, 2002; Roberts, Bonnici, Mackinnon, & Worcester, 2001; Worcester & Burns, 1975).

A Likert item in practice is considered symmetric or balanced when observations contain equal amounts of positive and negative positions. The "distance" between each successive Likert item is traditionally assumed to be equal—i.e., the psycho-metric distance between 1 and 2 is equidistant to 2–3. In terms of research ethics an equidistant presentation by the researcher is important; otherwise it will introduce a research bias into the analysis. A good Likert scale will present a symmetry of Likert items about a middle category that have clearly defined linguistic qualifiers for each item. In such symmetric scaling, equidistant attributes will typically be more clearly observed or, at least, inferred. It is when a Likert scale is symmetric and equidistant that it will behave like an interval-level measurement. Reips and Funke (2008) showed that interval-level measurement was better achieved by a visual analogue scale. Another perspective applies a polytomous Rasch model to infer that the Likert items are interval-level estimates on a continuum, and thus that statements reflect increasing levels of an attitude or trait—e.g., as might be used in grading in educational assessment, and scoring of performances by judges.

Any approximation suffers from information loss; specifying the magnitude and nature of that loss, though, can be challenging. Fortunately, information measures of sample adequacy have a long history. These, for example, have been articulated in the "information criterion" published in Akaike (1974) using information entropy. The Akaike information criterion (AIC) measures the information lost when a given model is used to describe population characteristics. It describes the trade-off between bias and variance (accuracy and complexity) of a model. Given a set of candidate models for the data, the preferred model is the one with the minimum AIC value (minimum information loss); it rewards goodness of fit while penalizing an increasing number of estimated parameters. The Schwarz criterion (Ludden, Beal, & Sheiner, 1994; Pauler, 1998) is closely related to AIC, and is sometimes called the Bayesian information criterion.

Ideally, responses to survey questions should yield discrete measurements that are dispersed and balanced—this maximizes the information contained in responses. Researchers would like respondents to make a definite choice rather than choose neutral or intermediate positions on a scale. Unfortunately, cultural, presentation, and subject matter idiosyncrasies can effectively sabotage this objective (Dietz, Bickel, & Scheffer, 2007; Lemmens, 2008; Sohn, 2005). Lee, Jones, Mineyama, and Zhang (2002) point out that Asian survey responses tend to be more closely compressed around the central point than Western responses; superficially, this suggests that Asian surveys may actually yield less information (dispersion) than Western surveys. To improve responses, some researchers suggest that a Likert scale without a midpoint would be preferable, provided it does not affect the validity or reliability of the responses. Cox (1980); Devasagayam (1999); H. H. Friedman and Amoo (1999); H. H. Friedman, Wilamowsky, and Friedman (1981); Komorita (1963); Komorita and Graham (1965); Matell and Jacoby (1972); and Wildt and Mazis (1978) have all demonstrated that as the number of scale steps is increased, respondents' use of the midpoint category decreases. Additionally, Clarke et al. (2002); Roberts et al. (2001); Worcester and Burns (1975); J. C. Chan (1991); Dawes (2012); Dawes, Riebe, and Giannopoulos (2002); H. H. Friedman et al. (1981); and Sparks, Desai, Thirumurthy, Kistenberg, and Krishnamurthy (2006) have found that grammatically balanced Likert scales are often unbalanced in interpretation. Worcester and Burns also concluded that a 4-point scale without a midpoint appears to push more respondents towards the positive end of the scale. The previously cited research concludes that Likert scales are subject to distortion from at least three causes. Subjects may:

1. Avoid using extreme response categories (central tendency bias)
2. Agree with statements as presented (acquiescence bias)
3. Try to portray themselves or their organization in a more favorable light (social desirability bias)

Designing a balanced Likert scale (with an equal number of positive and negative statements) can obviate the problem of acquiescence bias, since acquiescence on positively keyed items will balance acquiescence on negatively keyed items, but there are no widely accepted solutions to central tendency and social desirability biases.

The number of possible responses may matter as well. Likert used five ordered response levels, but seven and even nine levels are common as well. Allen and

Seaman (2007) concluded that a 5- or 7-point scale may produce slightly higher mean scores relative to the highest possible attainable score, compared to those produced from a 10-point scale, and this difference was statistically significant. In terms of the other data characteristics, there was very little difference among the scale formats in terms of variation about the mean, skewness, or kurtosis.

From another perspective, a Likert scale can be considered as a grouped form of a continuous scale. This is important in path analysis, since you implicitly treat the variable as if it were continuous for correlational analysis. Likert scales are clearly ordered category scales, as required for correlational work, but the debate among methodologists is whether they can be treated as equal interval scales.

When a Likert scale approximates an interval-level measurement, we can summarize the central tendency of responses using either the median or the mode, with "dispersion" measured by standard deviations, quartiles, or percentiles. Characteristics of the sample can be obtained from nonparametric tests such as chi-squared test, Mann–Whitney test, Wilcoxon signed-rank test, or Kruskal–Wallis test (J. Chan, Tan, & Tay, 2000; J. P. E. Chan, Tan, Tay, & Nanyang Technological University. School of Accountancy and Business, 2000; Jamieson, 2004; Norman, 2010).

Information clearly is lost in the mapping of beliefs to a Likert scale; how much information is lost is probably unknowable in practice. But the loss in information from that that would exist if our modeling assumptions (e.g., Gaussian beliefs) were actually true can be assessed. At this point, let us more precisely define the concepts of (1) informativeness, (2) bias, and (3) dispersion in Likert representations of survey subject belief distributions, starting with graphical depictions of bias and dispersion in the following figures (Figs. 7.3 and 7.4).

Proper instrument design requires a standardization of Likert responses so that ideally one standard deviation of the actual distribution of beliefs will shift the Likert score 1 point higher—this is comparable to the process of keeping the survey instrument "on scale" measuring beliefs in similar units to the subject's normal conventions. In addition, the mode of subject beliefs (i.e., what the largest number of people believe or agree upon) is presumed to center somewhere in the range 2 through 4 of the 5-point scale, with all other values being the "extremes"—response "1" or response "5." This is more or less what survey researchers aspire to, where the level of agreement or disagreement is measured (i.e., is "on scale") and the scaling is considered symmetric or "balanced" because there are equal amounts of positive and negative positions (A. C. Burns & Bush, 2005, 2000). Most of the weight of the Gaussian belief distribution should lie within the Likert range 2 through 4 of the 5-point scale. Survey researchers can credibly move the range around, but probably should not try to alter the subject beliefs if they are trying to conduct an objective survey.

Weaknesses in data can be effectively addressed by increasing the sample size. This works for multicollinear data, non-Gaussian data, and Likert data as well. But since data collection is costly, it is desirable to increase sample size as little as possible. The path analysis literature is surprisingly vague on how much of an increase is needed. Jöreskog (1971a, 1971b) suggests increases of two orders of magnitude,

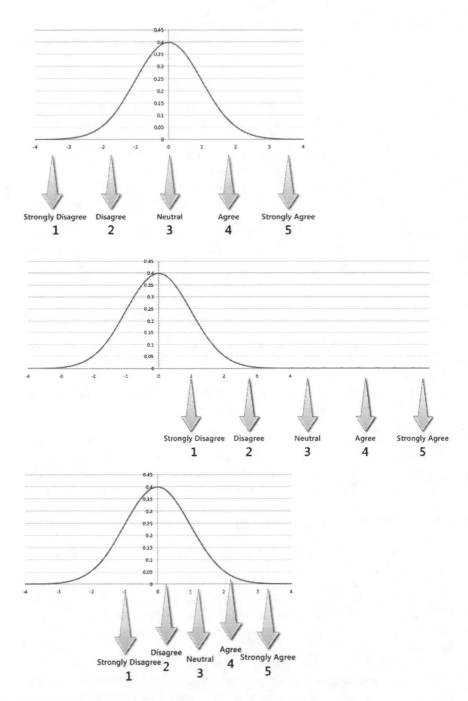

Fig. 7.3 Dispersion in balanced, unbalanced, and mis-scaled Likert mappings

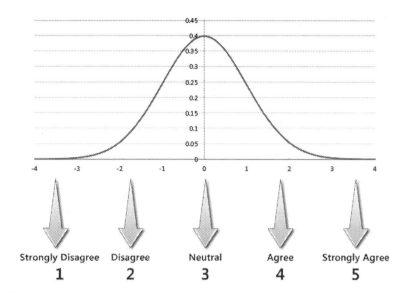

Fig. 7.4 Continuously varying strength of belief is approximated with 5 levels of Likert scale belief

but without offering causes or mitigating factors. If we assume that survey costs increase commensurately with sample size, then for most projects two orders of magnitude is likely to be prohibitive.

For example, in the path analysis approaches implemented in LISREL and AMOS software, for reasonably large samples, when the number of Likert categories is 4 or higher and skew and kurtosis are within normal limits, use of maximum likelihood is justified. In other cases some researchers use weighted least squares based on polychoric correlation. Jöreskog (1970a, 1970b, 1969, 1971a, 1971b, 1993) in Monte Carlo simulation found that phi, Spearman rank correlation, and Kendall tau-b correlation performed poorly whereas tetrachoric correlation with ordinal data such as Likert-scaled data was robust and yielded better fit.

7.3 Fisher Information in Likert Items

Fisher information (denoted here as $I_{\text{sample}}(\text{parameter})$) is additive; the information yielded by two independent samples is the sum of the separate sample's information: $I_{x,y}(\theta) = I_y(\theta) + I_x(\theta)$. Furthermore, the information in n independent sample observations is n times that in a single observation $I_n(\theta) = nI(\theta)$.

Assume that a survey collects n independent k-point Likert scale observations for each of the survey questions. Let the Likert scale represent a polytomous Rasch model (with say 5, 7, or 9 divisions) (Alphen, Halfens, Hasman, & Imbos, 2008;

T. Bond & Fox, 2007; Fitzpatrick et al., 2004; Norquist, Fitzpatrick, Dawson, & Jenkinson, 2004; White & Velozo, 2002). We can take the perspective of a polytomous Rasch model, assuming that the responses to the survey map an underlying Gaussian $N\left(\mu,\sigma^2\right)$ belief distribution to a Likert item across the population of subjects surveyed for a particular question on the survey.

Let $F\left(\cdot\,|\,\mu,\sigma\right)$ and $f\left(\cdot\,|\,\mu,\sigma\right)$ be cdf and pdf, respectively, of the underlying belief distribution. Presume that we use a metric scale that sets $\sigma^2=1$ (or alternately that the Likert "bin" partitions are spaced σ units apart). Let the Likert "bins" of the multinomial response distribution be the set $\left\{\left(x_1\in\left(-\infty,1\right]\right),\left\{x_i\in\left(i-1,i\right]\right\}_{i=2}^{k-1},\left(x_k\in\left(1,\infty\right)\right)\right\}$ where k is the total number of bins (usually 5, 7, or 9). Then the parameters $\{pi\}$ of the multinomial distribution of the "bin" summing of Likert items will be the set $\left\{p_1=F\left(1\,|\,\mu\right),\left\{p_i=F\left(i\,|\,\mu\right)-F\left(i-1\,|\,\mu\right)\right\}_{i=2}^{k-1},\ p_k=1-F\left(k-1\,|\,\mu\right)\right\}.$

A particular bin i is *filled* with probability of pi and *not filled* with probability $1-pi$; let n independent survey questions result in that bin being filled θi times, and not filled $n-\theta i$ times. If Bi is a logical variable that indicates whether the ith bin of the Likert item was chosen, then all possible outcomes for the Likert item can be represented $B_1\cup B_2\cup\ldots\cup B_{k-1}=\cup_{i=1}^{k-1}B_i$ since if none of the first $k-1$ bins were chosen, then the kth bin must have been chosen. Let the Fisher information in the ith bin of a sample of n Likert items be I_{B_i}. Since Bi is a logical variable, it can be perceived as a Bernoulli trial—a random variable with two possible outcomes, "success" with probability of pi and "failure," with probability of $1-pi$. The Fisher information contained in a sample of n independent Bernoulli trials for Bi where there are m successes, and where there are $n-m$ failures is

$$\mathbb{I}_{B_i}\left(p_i\right)=-E_{p_i}\left[\frac{\partial^2}{\partial p_i^{\,2}}\ln\left(f\left(m;p_i\right)\right)\right]=$$

$$-E_{p_i}\left[\frac{\partial^2}{\partial p_i^{\,2}}\ln\left(p_i^{\,m}\left(1-p_i\right)^{n-m}\frac{\left(m+\left(n-m\right)\right)!}{m!\left(n-m\right)!}\right)\right]=$$

$$=-E_{p_i}\left[\frac{\partial^2}{\partial p_i^{\,2}}\left(m\ln\left(p_i\right)+\left(n-m\right)\ln\left(1-p_i\right)\right)\right]=-E_{p_i}\left[\frac{\partial}{\partial p_i}\left(\left(\frac{m}{p_i}+\frac{\left(n-m\right)}{1-p_i}\right)\right)\right]=$$

$$=-E_{p_i}\left[\left(\frac{m}{p_i^{\,2}}+\frac{\left(n-m\right)}{\left(1-p_i\right)^2}\right)\right]=\left(\frac{np_i}{p_i^{\,2}}+\frac{n\left(1-p_i\right)}{\left(1-p_i\right)^2}\right)=\frac{n}{p_i\left(1-p_i\right)}$$

This is the reciprocal of the variance of the mean number of successes in n Bernoulli trials, as expected. The Fisher information contained in a sample of n independent Bernoulli trials for all possible outcomes for n Likert items $\cup_{i=1}^{k-1}B_i$ is

$$I_{\cup_{i=1}^{k-1}B_i}=\sum_{i=1}^{k-1}\left(\frac{n}{p_i\left(1-p_i\right)}\right)$$

Compare this to the Fisher information in a sample of n observations from a Gaussian $N(\mu,\sigma^2)$ belief distribution, which is estimated $\hat{I}_n = \dfrac{n}{\sigma^2}$ (and which incidentally is independent of location parameter μ as $\hat{I}_n$ is the inverse of the variance). Then estimator $\hat{\omega}$ can be computed as the ratio of information content in these two different mappings from the same survey sample:

$$\hat{\omega} \triangleq \frac{\dfrac{n}{\sigma^2}}{\displaystyle\sum_{i=1}^{k-1}\left(\frac{n}{p_i(1-p_i)}\right)} = \frac{1}{\sigma^2 \displaystyle\sum_{i=1}^{k-1}\left(\frac{1}{p_i(1-p_i)}\right)}$$

Thus the lower bound on a sample that uses a Likert mapping will need to be $\hat{\omega}$ times as large as one that assumes a full Gaussian belief distribution. A Likert scale mapping of what is an inherently continuous distribution of beliefs in the population results in a significant increase in the sample size needed for estimation—by a factor of at least two orders of magnitude (i.e., 100 times) (Figs. 7.5 and 7.6).

There are three things that should be noted concerning multiplier for sample size estimates for processing Likert data when an assumption of Gaussian data has been made in the data analysis:

First, any difference of the actual sample standard deviation from the equidistant scale of the Likert items requires larger sample sizes; but the minimum sample size for any Likert-mapped dataset will be 100 times as large as that that would

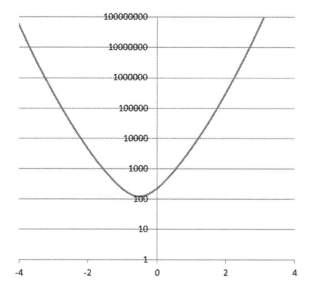

Fig. 7.5 Unbalanced Likert mappings cause sample size increases by the multiple shown on the y-axis when the central category on the Likert mapping is biased by the number of standard deviations shown on the x-axis

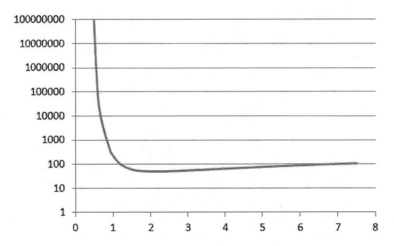

Fig. 7.6 Mis-scaled Likert mappings cause sample size increases by the multiple shown on the *y*-axis when the standard deviation of the actual belief distribution is mis-scaled by number of Likert scale intervals shown on the *x*-axis

be required if you had all of the original information in the Gaussian distribution of beliefs. The information loss from using Likert scaling is at least two orders of magnitude—the increase in sample size is at least two orders of magnitude.

Second, the sample is most informative when location of the Gaussian mean coincides with the central Likert bin. This emphasizes the importance of "balanced" designs for the Likert scaling in the survey instrument.

Third, information in the underlying belief distribution, which has a support, does not depend on the mean of an assumed underlying Gaussian distribution of data. The Likert mapping information content does depend on the mean and is sensitive to the Likert scale being "balanced"—this is controlled in the survey design.

7.4 Examples of Likert Scalings and Their Consequences

In order to gain a more intuitive understanding of how the metric in this paper functions in comparison with the Srinivasan and Basu (1989) metric, we can operationalize the Likert mapping as a survey instrument $T \otimes S$ that "bins" Y values (i.e., the measure of unobservable underlying phenomenon X) into responses Z on a 5-point scale: $T : X \rightarrow \left(Y + \theta \right)$ and $S : Y \rightarrow \left(Z + \delta \right)$. Random variable $\hat{\varepsilon}$ describes the error (information loss) in the mapping of survey instrument $T \otimes S : X \rightarrow \left(Z + \varepsilon \right)$. Conceptually, $\hat{\varepsilon} = \hat{\delta} + \hat{\theta}$ where $\hat{\theta}$ is the part resulting from misspecification of the survey instrument (bias and dispersion) and $\hat{\delta}$ is the part resulting from approximating a continuous variable into the five bins in the Likert scale. The seven sets of responses, including a restatement of Anscombe's quartet which we previously encountered in chapter 3 on PLS-PA, encapsulate several challenges—skewness, kurtosis, outliers, non-informative data, and a nonlinear (parabolic) data (Fig. 7.7; Table 7.1).

Srinivasan and Basu (1989) evaluated the information content of Likert data item Z (an m-point Likert scale variable) that approximates some continuous variable $\hat{Y}$ that in turn approximates some unobservable belief or phenomenon that the researcher wishes to measure. They assumed that $\hat{Y}$ is composed of a true "score" $\hat{X}$ and error $\hat{\varepsilon}$ (which in their formulation is additive, but which we will allow to take on more complex functional forms). Then in their formulation

$$\hat{Y} = \hat{X} + \hat{\varepsilon} \text{ where } \hat{X} \sim N\left(0, 1\right) \text{ and } \hat{\varepsilon} \sim N\left(0, \theta^2\right) \text{ and } \rho\left(\hat{X}, \hat{\varepsilon}\right) = 0.$$

Thus $\hat{Y} \sim N\left(0, 1 + \theta^2\right)$ and using the fact that the Pearson correlation coefficient is invariant (up to a sign) to separate changes in location and scale in the two variables, we

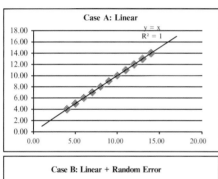

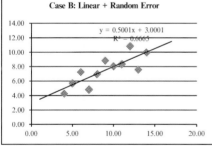

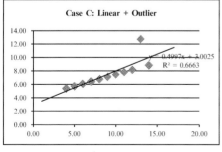

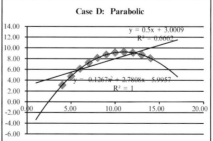

Fig. 7.7 $\left\{\hat{X} \times \hat{Y}\right\}$: four datasets with $R^2 = 0.666$ and three datasets with $R^2 = 1$

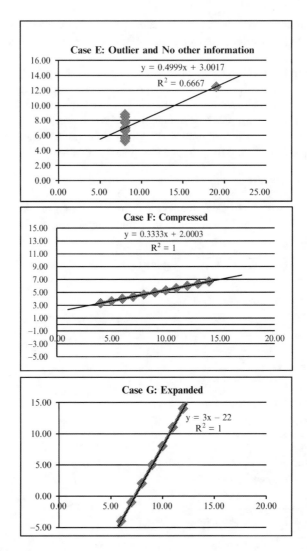

Fig. 7.7 (continued)

can recompute the Srinivasan and Basu (1989) information metric $I_z = \dfrac{\rho^2(Z,X)}{\rho^2(Y,X)}$.

Since the correlation $\rho(\hat{X},\hat{Y}) = \rho(\hat{X}+\hat{\varepsilon},\hat{X}) = \rho(\hat{X},\hat{X}) + \rho(\hat{\varepsilon},\hat{X}) = 1+0 = 1$ this

metric is identically $I_z \equiv \rho^2(\hat{Z},\hat{X})$.

Table 7.1 $\{\hat{X} \times \hat{Y}\}$: four datasets with $R^2 = 0.666$ and three datasets with $R^2 = 1$; binned into Likert variable Ž

Observation	Case A Bin 1, 2, 3x	Case B 1x	1y	Bin 1y 1z	Case C 2x	2y	Bin 2y 2z	Case D 3x	3y	Bin 3y 3z
1	10.00	10.00	8.04	10.00	10.00	9.14	10.00	10.00	7.46	8.00
2	8.00	8.00	6.95	8.00	8.00	8.14	10.00	8.00	6.77	8.00
3	10.00	13.00	7.58	8.00	13.00	8.74	10.00	13.00	12.74	10.00
4	10.00	9.00	8.81	10.00	9.00	8.77	10.00	9.00	7.11	8.00
5	10.00	11.00	8.33	10.00	11.00	9.26	10.00	11.00	7.81	8.00
6	10.00	14.00	9.96	10.00	14.00	8.10	10.00	14.00	8.84	10.00
7	6.00	6.00	7.24	8.00	6.00	6.13	8.00	6.00	6.08	8.00
8	4.00	4.00	4.26	6.00	4.00	3.10	4.00	4.00	5.39	6.00
9	10.00	12.00	10.84	10.00	12.00	9.13	10.00	12.00	8.15	10.00
10	8.00	7.00	4.82	6.00	7.00	7.26	8.00	7.00	6.42	8.00
11	6.00	5.00	5.68	6.00	5.00	4.74	6.00	5.00	5.73	6.00
Mean	8.36	9.00	7.50	8.36	9.00	7.50	8.73	9.00	7.50	8.18
Std dev	2.15744	3.316625	2.031568	1.747726	3.316625	2.031657	2.053821	3.316625	2.030424	1.401298
Skewness	-1.01393	-8.1E-17	-0.06504	-0.40869	-8.1E-17	-1.3158	-1.58382	-8.1E-17	1.855495	-0.12334
Kurtosis	-0.20589	-1.2	-0.5349	-1.62132	-1.2	0.846123	1.743956	-1.2	4.384089	-0.45267

(continued)

Table 7.1 (continued)

Observation	Case E			Case F			Case G		
	4x	4y	4z	5x	5y	5z	6x	6y	6z
1	8.00	6.58	8.00	10.00	8.00	8.00	10.00	5.33	6.00
2	8.00	5.76	6.00	8.00	2.00	2.00	8.00	4.67	6.00
3	8.00	7.71	8.00	13.00	17.00	10.00	13.00	6.33	8.00
4	8.00	8.84	10.00	9.00	5.00	6.00	9.00	5.00	6.00
5	8.00	8.47	10.00	11.00	11.00	10.00	11.00	5.67	6.00
6	8.00	7.04	8.00	14.00	20.00	10.00	14.00	6.67	8.00
7	8.00	5.25	6.00	6.00	-4.00	2.00	6.00	4.00	6.00
8	19.00	12.50	10.00	4.00	-10.00	2.00	4.00	3.33	4.00
9	8.00	5.56	6.00	12.00	14.00	10.00	12.00	6.00	6.00
10	8.00	7.91	8.00	7.00	-1.00	2.00	7.00	4.33	6.00
11	8.00	6.89	8.00	5.00	-7.00	2.00	5.00	3.67	4.00
Mean	9.00	7.50	8.00	9.00	5.00	5.82	9.00	5.00	6.00
Std dev	3.316625	2.030579	1.549193	3.316625	9.949874	3.842348	3.316625	1.105431	1.264911
Skew	3.316625	1.506818	0	-8.1E-17	-8.1E-17	0.052992	-8.1E-17	2.96E-15	0
Kurt	11	3.151315	-1.11111	-1.2	-1.2	-2.22488	-1.2	-1.2	0.416667

A more general formulation allows $\rho\left(\hat{X},\hat{Y}\right)\in\left[-1,1\right]$. In case A, F, and G $\rho\left(\hat{X},\hat{Y}\right)=1$ as assumed in Srinivasan and Basu's formulation; in cases B, C, D, E $\rho\left(\hat{X},\hat{Y}\right)=0.666$ (Table 7.2).

The table suggests that there are two significant weaknesses of the Srinivasan and Basu (1989) metric IZ:

1. The value often converges outside the purported [0,1] range of the statistic (as in C, D, F, and G).
2. Even small changes in survey setup, or of question interpretation by subjects, can have a huge impact on reported information content.

The Fisher information statistic does not have a value when $R^2 = 1$, but otherwise converges to values that are intuitive in the sense that they suggest that information captured from subjects is fairly stable.

7.5   Design Trade-Offs in Likert Scaling

There are specific design trade-offs which must be made in asserting that a particular Likert sample contains a specific amount of information concerning the specific research question that the survey is designed to answer. This chapter's examples using Gaussian beliefs showed that the Fisher information metric is more informative, stable, and reliable than previous approaches to assessing adequacy of survey datasets. Such metrics also accentuate the importance of balanced survey design, potentially without a midpoint, as suggested by Cox (1980), Devasagayam (1999), H. H. Friedman and Amoo (1999), H. H. Friedman et al. (1981), Komorita (1963), Komorita and Graham (1965), Matell and Jacoby (1972), and Wildt and Mazis (1978). It also suggests that where grammatically balanced Likert scales are unbalanced in interpretation, the impact on survey conclusions may be significant (Clarke et al., 2002; Roberts et al., 2001; Worcester & Burns, 1975; J. C. Chan, 1991; Dawes, 2012; Dawes et al., 2002; H. H. Friedman et al., 1981; Sparks et al., 2006). The previous arguments suggest that any difference of the actual sample standard deviation from the equidistant scale of the Likert items requires larger sample sizes; but the minimum sample size for any Likert-mapped dataset will be orders of magnitute larger than that would be required if you had all of the original information in the Gaussian distribution of beliefs. Additionally, the sample is most informative when location of the Gaussian mean coincides with the central Likert bin. This emphasizes the importance of "balanced" designs for the Likert scaling in the survey instrument. Finally, information in the underlying belief distribution, which has a support, does not depend on the mean of an assumed underlying Gaussian distribution of data. The Likert mapping information content does depend on the mean and is sensitive to the Likert scale being "balanced."

Additionally, the example identified a practical issue in the implementation of the Srinivasan and Basu (1989) metric IZ, in that its value often converges outside

Table 7.2 Information content of seven cases

	A	B	C	D	E	F	G
$I_z = \dfrac{\rho^2(Z,X)}{\rho^2(Y,X)}$	0.894427	0.864447	1.014439	1.22563	0.274983	na	na
$\hat{\omega} \triangleq \dfrac{\frac{n}{\sigma^2}}{\sum_{i=1}^{k-1}\left(\frac{n}{p_i(1-p_i)}\right)} = \dfrac{1}{\sigma^2 \sum_{i=1}^{k-1}\left(\frac{1}{p_i(1-p_i)}\right)}$	0.144737	0.141026	0.166667	0.152778	0.141026	na	na
Info per observation $= \sum_{i=1}^{S-1}\left(\dfrac{1}{p_i(1-p_i)}\right)$	0.628099	0.644628	0.545455	0.595041	0.644628	0.628099	0.528926
$\text{FI} = \sum_{i=1}^{S-1}\left(\dfrac{n}{p_i(1-p_i)}\right)$	6.909089	7.090908	6.000005	6.545451	7.090908	6.909089	5.818186

the purported [0,1] range of the statistic (as in examples C, D, F, and G) and that even small changes in survey setup, or of question interpretation by subjects, can have a huge impact on information content reported by metric *IZ*.

In contrast, the Fisher information estimator $\hat{\omega}$ only fails to compute in the limiting case where $R^2 = 1$, but otherwise converges to values that are intuitive in the sense that they suggest that information captured from subjects is fairly stable. This reinforces conclusions rendered by Lee et al. (2002) concerning survey designs across cultures.

The assumption of Gaussian belief distributions may or may not be justified in practice. Studies by Brandstätter, Kühberger, and Schneider (2002); Kühberger (1995, 1998); and Kühberger, Schulte-Mecklenbeck, and Perner (1999, 2002) have concluded that people do not generally hold strong, stable, and rational beliefs, and that their responses are very much influenced by the way in which decisions are framed. This would tend to indicate that some distribution besides Gaussian distributions would be most appropriate for human beliefs. Nonetheless, the Gaussian assumption is widely used, especially in survey research using tools such as AMOS or LISREL. And this assumption, and the amount of information loss $\hat{\omega}$, implies that sample sizes need to increase to offset the mapping loss.

An alternative approach to assessing the informativeness of Likert items, with significantly reduced demands on sample size, could invoke Bayesian conjugate families of distributions. Such an approach would essentially pool prior research findings (potentially both qualitative and quantitative) in the prior distribution, with a likelihood function built from the data. Given the categorical nature of Likert mappings, a multinomial-Dirichlet conjugate family of distributions would be appropriate for Bayesian analysis of Likert survey data. Such approaches have been explored in the artificial intelligence and quality control fields (Dietz et al., 2007; Lemmens, 2008; Sohn, 2005) and the statistics developed in Crook and Good (1980) and Good (1976). So far, the multinomial-Dirichlet conjugate family of distributions appears not to have been applied to the analysis of survey-generated Likert data.

7.6   Known Unknowns: What Is a Latent Variable?

What do we really know about our unmeasurable, so-called latent variables? We have defined them as constructs that we think are real, but which cannot be directly measured. In a social context, these may be abstractions like trust, intent, and happiness. In other cases latent variables are cultural—in some cultures, trust may only apply to family; in others, it might be a distinguishing feature of national culture. Latent variables might not even be real—rather they could be shared perceptions, possibly even fantasies. For example, one survey concluded that 77 % of adult Americans believe in angels (Johnson, 2011). Thus "angels" appear to be appropriate latent constructs. If you are not comfortable trying to find indicator variables for a structural model of latent variables germane to angels, then consider the results of another poll—84 % of children believe in Santa Claus. We could construct some structural relationships between Santa's reindeer, aerodynamics, and overall mass—all unobservable—and build a measurement model on subjects' perceptions of Santa. Given the high rate of belief in Santa's existence, we are likely to experience high response to any survey we would construct. In both cases, the measured variables would be perceptions, since it would be difficult to acquire physical evidence of either angels or Santa.

Whatever the basis in reality is for one's structural model, its implementation will involve a linear combination of measured factors. We see this elsewhere in statistics, in contrasts: linear combinations of two or more factor level means whose coefficients add up to zero. In SEM, clustering of measured factors around latent variables is decided a priori by the researcher as a part of model specification, perhaps based on pretests and principal component analysis.

From a practical standpoint, the structural or inner model is merely an artifact that identifies our model as a structural equation model. We could just as easily substitute a linear combination of measured factors everywhere that a latent variable appears in the model. And thus from a practical standpoint, any latent variable structural equation model has an equivalent linear model that is constructed entirely of observed variables.

References

Akaike, H. (1974). A new look at the statistical model identification. *IEEE Transactions on Automatic Control, 19*(6), 716–723.

Allen, I. E., & Seaman, C. A. (2007). Likert scales and data analyses. *Quality Progress, 40*(7), 64–65.

Alphen, A., Halfens, R., Hasman, A., & Imbos, T. (2008). Likert or Rasch? Nothing is more applicable than good theory. *Journal of Advanced Nursing, 20*(1), 196–201.

Basmann, R. L. (1963). The causal interpretation of non-triangular systems of economic relations. *Econometrica, 31*, 439–448.

Bond, T., & Fox, C. (2007). *Applying the Rasch model: fundamental measurement in the human sciences.* Mahwah, NJ: Lawrence Erlbaum.

Brandstätter, E., Kühberger, A., & Schneider, F. (2002). A cognitive-emotional account of the shape of the probability weighting function. *Journal of Behavioral Decision Making, 15*(2), 79–100.

Burns, A. C., & Bush, R. F. (2000). *Marketing research* (Globalization, Vol. 1, p. 7). Upper Saddle River, NJ: Prentice Hall.

Burns, A. C., & Bush, R. F. (2005). *Basic marketing research: using Microsoft Excel data analysis.* Upper Saddle River, NJ: Prentice Hall.

Chan, J. C. (1991). Response-order effects in Likert-type scales. *Educational and Psychological Measurement, 51*(3), 531–540.

Chan, J. P. E., Tan, K. H. C., Tay, K. Y., & Nanyang Technological University. School of Accountancy and Business. (2000a). *The impact of electronic commerce on the role of the middleman.* Singapore: School of Accountancy and Business Nanyang Technological University.

Chan, J. P. E., Tan, K. H. C., & Tay, K. Y. (2000b). *Likert scales in interval measurement.* Singapore: Nanyang Technological University.

Clarke, S., Worcester, J., Dunlap, G., Murray, M., & Bradley-Klug, K. (2002). Using Multiple Measures to Evaluate Positive Behavior Support. A Case Example. *Journal of Positive Behavior Interventions, 4*(3), 131–145.

Cox, E. P., III. (1980). The optimal number of response alternatives for a scale: a review. *Journal of Marketing Research, 17*, 407–422.

Crook, J. F., & Good, I. J. (1980). On the application of symmetric Dirichlet distributions and their mixtures to contingency tables, Part II. *The Annals of Statistics, 8*(6), 1198–1218.

Dawes, J. (2012). Do data characteristics change according to the number of scale points used? An experiment using 5 point, 7 point and 10 point scales. *International Journal of Market Research, 50*, 66.

Dawes, J., Riebe, E., & Giannopoulos, A. (2002). The impact of different scale anchors on responses to the verbal probability scale. *Canadian Journal of Marketing Research, 20*(1), 77–80.

Devasagayam, P. R. (1999). *The effects of randomised scales on scale checking styles and reaction time.* Paper presented at the 1999 Marketing Management Association Conference Proceedings.

Dietz, L., Bickel, S., & Scheffer, T. (2007). *Unsupervised prediction of citation influences.* Paper presented at the Proceedings of the 24th international conference on Machine learning.

Fitzpatrick, R., Norquist, J. M., Jenkinson, C., Reeves, B. C., Morris, R. W., Murray, D. W., & Gregg, P. J. (2004). A comparison of Rasch with Likert scoring to discriminate between patients' evaluations of total hip replacement surgery. *Quality of Life Research, 13*(2), 331–338.

Friedman, H. H., & Amoo, T. (1999). Rating the rating scales. *Journal of Marketing Management, 9*(3), 114–123.

Friedman, J., Hastie, T., Rosset, S., Tibshirani, R., & Zhu, J. (2004). Discussion of boosting papers. *The Annals of Statistics, 32*, 102–107.

Friedman, J., Hastie, T., & Tibshirani, R. (2010). Regularization paths for generalized linear models via coordinate descent. *Journal of Statistical Software, 33*(1), 1.

Friedman, H. H., Wilamowsky, Y., & Friedman, L. W. (1981). A comparison of balanced and unbalanced rating scales. *The Mid-Atlantic Journal of Business, 19*(2), 1–7.

Good, I. J. (1976). On the application of symmetric Dirichlet distributions and their mixtures to contingency tables. *The Annals of Statistics, 4*, 1159–1189.

Gurevich, M. (1961). *The social structure of acquaintanceship networks.* Cambridge, MA: MIT.

Hill, T. P. (1995). A statistical derivation of the significant-digit law. *Statistical Science, 10*(4), 354–363.

Jamieson, S. (2004). Likert scales: how to (ab) use them. *Medical Education, 38*(12), 1217–1218.

Johnson, J. (2011). 77 Percent of American's believe in angels. Retrieved on Jan 3, 2012, from http://www.inquisitr.com/171741/77-percent-of-americans-believe-in-angels-polling/

Jöreskog, K. G. (1969). A general approach to confirmatory maximum likelihood factor analysis. *Psychometrika, 34*(2), 183–202.

Jöreskog, K. G. (1970a). A general method for estimating a linear structural equation system. In *Structural equation models in the social sciences.* New York, NY: Seminar press.

Jöreskog, K. G. (1970b). A general method for analysis of covariance structures. *Biometrika, 57*(2), 239–251.

Jöreskog, K. G. (1971a). Simultaneous factor analysis in several populations. *Psychometrika, 36*(4), 409–426.

Jöreskog, K. G. (1971b). Statistical analysis of sets of congeneric tests. *Psychometrika, 36*(2), 109–133.

Jöreskog, K. G. (1993). Testing structural equation models. In *Sage focus editions* (Vol. 154, p. 294). Thousand Oaks, CA: Sage.

Komorita, S. S. (1963). Attitude content, intensity, and the neutral point on a Likert scale. *The Journal of Social Psychology, 61*(2), 327–334.

Komorita, S. S., & Graham, W. K. (1965). Number of scale points and the reliability of scales. *Educational and Psychological Measurement, 25*(4), 987–995.

Kühberger, A. (1995). The framing of decisions: a new look at old problems. *Organizational Behavior and Human Decision Processes, 62*(2), 230–240.

Kühberger, A. (1998). The influence of framing on risky decisions: a meta-analysis. *Organizational Behavior and Human Decision Processes, 75*(1), 23–55.

Kühberger, A., Schulte-Mecklenbeck, M., & Perner, J. (1999). The effects of framing, reflection, probability, and payoff on risk preference in choice tasks. *Organizational Behavior and Human Decision Processes, 78*(3), 204–231.

Kühberger, A., Schulte-Mecklenbeck, M., & Perner, J. (2002). Framing decisions: hypothetical and real. *Organizational Behavior and Human Decision Processes, 89*(2), 1162–1175.

Lee, J. W., Jones, P. S., Mineyama, Y., & Zhang, X. E. (2002). Cultural differences in responses to a Likert scale. *Research in Nursing & Health, 25*(4), 295–306.

Lemmens, L. F. (2008). *Quality control in small groups.* Paper presented at the AIP Conference Proceedings.

Likert, R. (1974). The method of constructing an attitude scale. In *Scaling: a sourcebook for behavioral scientists* (pp. 233–243). Chicago, IL: Aldine.

Likert, R., Roslow, S., & Murphy, G. (1934). A simple and reliable method of scoring the Thurstone attitude scales. *The Journal of Social Psychology, 5*(2), 228–238.

Ludden, T. M., Beal, S. L., & Sheiner, L. B. (1994). Comparison of the Akaike Information Criterion, the Schwarz criterion and the F test as guides to model selection. *Journal of Pharmacokinetics and Pharmacodynamics, 22*(5), 431–445.

Mandelbrot, B. B. (1982). *The fractal geometry of nature.* New York, NY: Times Books.

Matell, M. S., & Jacoby, J. (1972). Is there an optimal number of alternatives for Likert-scale items? Effects of testing time and scale properties. *Journal of Applied Psychology, 56*(6), 506.

Norman, G. (2010). Likert scales, levels of measurement and the "laws" of statistics. *Advances in Health Sciences Education, 15*(5), 625–632.

Norquist, J. M., Fitzpatrick, R., Dawson, J., & Jenkinson, C. (2004). Comparing alternative Rasch-based methods vs raw scores in measuring change in health. *Medical Care, 42*(1), 1.

Pauler, D. K. (1998). The Schwarz criterion and related methods for normal linear models. *Biometrika, 85*(1), 13–27.

Reips, U. D., & Funke, F. (2008). Interval-level measurement with visual analogue scales in Internet-based research: VAS generator. *Behavior Research Methods, 40*(3), 699–704.

Roberts, S. B., Bonnici, D. M., Mackinnon, A. J., & Worcester, M. C. (2001). Psychometric evaluation of the Hospital Anxiety and Depression Scale (HADS) among female cardiac patients. *British Journal of Health Psychology, 6*(4), 373–383.

Sohn, S. Y. (2005). Random effects logistic regression model for ranking efficiency in data envelopment analysis. *Journal of the Operational Research Society, 57*(11), 1289–1299.

Sparks, R., Desai, N., Thirumurthy, P., Kistenberg, C., & Krishnamurthy, S. (2006). *Measuring e-Commerce Satisfaction: Reward Error and the Emergence of Micro-Surveys.* Paper presented at the IADIS International e-Commerce Conference Proceedings.

Srinivasan, V., & Basu, A. K. (1989). The metric quality of ordered categorical data. *Marketing Science, 8*(3), 205–230.

Sterne, J. A. C., Smith, G. D., & Cox, D. R. (2001). Sifting the evidence—what's wrong with significance tests? Another comment on the role of statistical methods. *BMJ, 322*(7280), 226–231.

Westland, J. C. (2010). Lower bounds on sample size in structural equation modeling. *Electronic Commerce Research and Applications, 9*(6), 476–487.

White, L. J., & Velozo, C. A. (2002). The use of Rasch measurement to improve the Oswestry classification scheme. *Archives of Physical Medicine and Rehabilitation, 83*(6), 822–831.

Wildt, A. R., & Mazis, M. B. (1978). Determinants of scale response: label versus position. *Journal of Marketing Research, 15*, 261–267.

Worcester, R. M., & Burns, T. R. (1975). A statistical examination of the relative precision of verbal scales. *Journal of the Market Research Society, 17*(3), 181–197.

Chapter 8
Research Structure and Paradigms

John Ioannidis, a highly respected medical researcher, has a serious bone to pick—with the modern practitioners of the Galilean hypothetico-deductive model-data duality discussed in Chap. 6. Ioannidis' 2005 paper, provocatively titled "Why Most Published Research Findings Are False," has been the most downloaded technical paper in *PLoS Medicine* and one of the single most cited and downloaded papers in the past decade (J. P. A. Ioannidis, 2005; Ioannidis et al., 2001). In it, Ioannidis analyzed "49 of the most highly regarded research findings in medicine over the previous 13 years" comparing them with data from subsequent studies with larger sample sizes. His findings included the following: 7 (16 %) of the original studies were contradicted, 7 (16 %) of the effects were smaller than in the initial study, 20 (44 %) were replicated, and 11 (24 %) of the studies remained largely unchallenged (D. H. Freedman, 2010; J. P. A. Ioannidis, 2005; Liberati et al., 2009; McCarthy et al., 2008).

Ioannidis' "most research findings are false" assertion was not hyperbole; indeed only 44 % of these highly regarded findings could be replicated. His research was surprising and influential, and resulted in subsequent changes in the conduct of US clinical trials. Ioannidis called this failure to replicate findings the "Proteus phenomenon."

Weak research is often driven by an incentive to publish quickly, for fame, reputation, patent rights, or ability to publish results at all. Research priority is the credit given to the individual or group of individuals who first made a discovery or propose a theory. Priority debates have defined the form and context of modern science; yet as Stephen Jay Gould once remarked "debates about the priority of ideas are usually among the most misdirected in the history of science" (Gould, 1977).

Priority has been at the center of Western research traditions for four centuries. It is the primary reason that research journals exist. The early research journal *Philosophical Transactions of the Royal Society* was founded in the seventeenth century at a time that scientists did not publish; rather they competed in contests for employment. At that time, the act of publishing academic inquiry was similar to

© Springer International Publishing Switzerland 2015

139

J.C. Westland, *Structural Equation Models*, Studies in Systems,
Decision and Control 22, DOI 10.1007/978-3-319-16507-3_8

distribution of open-source software today: difficult to justify because of the lack of financial incentives. Yet it was highly effective in resolving priority disputes. Studies found that 92 % of cases of simultaneous discovery in the seventeenth century ended in priority dispute; this dropped to 72 % in the eighteenth century, 59 % by the latter half of the nineteenth century, and 33 % by the first half of the twentieth century (R. K. Merton, 1957, 1968a). That is not to say that publishers necessarily got priority right. A cynical but widely accepted view is called *Stigler's law of eponymy*: no scientific discovery is named after its original discoverer. Stigler drolly named the sociologist Robert K. Merton as the discoverer of Stigler's law to avoid contradiction (Stigler, 1980) referring to Merton's "Matthew Effect" (R. K. Merton, 1968a, 1988, 1995) where the rich get richer, the powerful more powerful, and the poor more destitute.

The Proteus phenomenon is less of an issue in the social sciences, only because it is virtually impossible to replicate the quasi-experiments that are the norm in the social sciences. Consequently, social science research findings are not subject to the same intense (and well-funded) scrutiny of medicine. That doesn't mean that they don't suffer from their own Proteus phenomenon. Since problems are unlikely to be detected after publication, control over social science research protocols must happen earlier in the process—at the time the research is designed. This chapter addresses the problems and potential controls over social sciences' own Proteus phenomenon.

8.1   The Quest for Truth

Statisticians like to think of their craft as a game, pitting them against nature, which keeps secret the true state of thing. Answering questions gives statisticians insight into the "true state of nature"—into the real world. Philosophers have been seeking the truth throughout much of the world's history. Some of the salient theories to arise from this quest have been the following:

1. *Correspondence theories* claim that true beliefs and true statements correspond to the actual state of affairs.
2. *Coherence theories* require a proper fit of elements within a whole system as a basis for asserting truth.
3. *Social constructivism* holds that truth is constructed by social processes, is historically and culturally specific, and is in part shaped through the power struggles within a community. Constructivism views all of our knowledge as "constructed," because it does not reflect any external "transcendent" realities.
4. *Consensus theory* holds that truth is whatever is agreed upon, or, in some versions, might come to be agreed upon, by some specified group.
5. *Pragmatic theory* articulated by the psychologist William James (R. K. Merton, 1968b) suggests that "The 'true' is only the expedient in our way of thinking, just as the 'right' is only the expedient in our way of behaving."
6. *Minimalist (deflationary) theories* reject the thesis that the concept or term *truth* refers to a real property of sentences or propositions.

The exact meaning of "*truth*" is open to wide interpretation requiring strong arguments. This may require more energy than statisticians may be willing to give up to the philosophical portion of their projects. Most are satisfied with G.E.P. Box's dictum that "All models are wrong, but some are useful" (Box & Draper, 2007). This chapter takes Box's (and indirectly James') "pragmatic" approach to the truth.

8.2 Research Questions

No matter what the topic is, the most important decision facing the researcher is choice of research question. This choice determines the data collected, method of analysis used, and ultimate meaning and utility of any answer that research might provide.

Data acquisition often is the costliest part of any project, and there is a natural tendency to look for questions that data can answer. This is understandable, and it can work if the researcher is honest in seeking a question to answer. Datasets are often quite limited by design in the amount and quality of information they contain, simply because of the cost of trade-off. Exploratory data analysis is directed towards finding out what information is contained in a database—and thus what questions can be answered. Research questions are dependent on dataset information—you cannot ask a research question of a database that it is unprepared to answer (no matter how much you torture the data).

8.3 Models

A model is a theoretical construct that represents something, with a set of variables and a set of logical and quantifiable relationships between them. Models are constructed to enable reasoning within an *idealized* logical framework about these processes; they are an important component of scientific inference and deduction.

When we use the term "idealized," we mean that the model can make explicit assumptions that are known to be false (or incomplete) in some detail. Such assumptions may be justified on the grounds that they simplify the model, while at the same time allowing the production of acceptably accurate solutions.

Another perspective would be that make these false, incomplete, simplifying assumptions knowing that they will produce errors, and the trade-off is with the quantifiable inaccuracy, resolution, or granularity of the errors in our conclusions. Tweaking these assumptions in subsequent research facilitates a stepping-stone approach to research, just as we might pause to catch our balance at each stepping stone in crossing a river (rather than trying to cross in one go). This latter perspective assumes that a particular research project is embedded in a more comprehensive research program.

Research programs are inclined to adopt the sort of new *venture options approaches* that we see in investment and industry—and for a similar end: to use scarce research time and funding in the most efficient way possible.

Venture options approaches parcel work out in increasing *quantities* of time and research funding. They adopt a tiered series of projects, often involving three steps of funding:

1. Proof of concept or pretest
2. Limited testing
3. Full set of tests

There are likely to be *qualitative* differences in these tests as well. They will fall into three categories:

1. Positioning options: These tie down the most useful and efficient set of assumptions for the models used in the full set of tests.
2. Scouting options: Former Secretary of Defense Donald Rumsfeld once famously remarked that there are "known unknowns, and unknown unknowns." Scouting options are designed to bring the latter "unknowns" into the realm of the "knowns."
3. Stepping-stone options: These lay out a stepwise approach to model specification, allowing us to collect small datasets to test the applicability of assumptions while we are still deciding the final form of the model. Exploratory analysis statistical techniques are typically very useful in analysis with a stepping-stone option approach.

8.4   Theory Building and Hypotheses

Differing objectives and traditions are likely to be associated with researchers using specific tools, or focusing on specific tasks within the hypothetico-deductive-inductive cycle of scientific inquiry. It is probably the work a researcher specializes in, as opposed to fundamentally differing philosophical bents, that dictates which tradition that researcher will choose to follow. Table 8.1 summarizes the objectives encompassed by various types of research.

This table looks at these objectives from a perspective more suitable to defining research disciplines. The character of available data and its inherent observability often determine definability of model constructs. Important factors here are the manner in which we (1) explore available data in search of a model *specification*; (2) find out how well a model derived from existing theory is *confirmed* by a particular dataset; (3) *discriminate* one model from another by determining which one is better supported by the data; or (4) *predict* the existence and *causal* direction of potential relationships between candidate factors. Such analyses are, respectively, referred to as (1) specification search, or *exploratory*; (2) *confirmatory*; (3) *discriminant*, and (4) *causal-predictive* or just plain *predictive* (Table 8.2).

Though researchers may specialize, a full research program will likely incorporate interpretive, descriptive, and positive phases. The appeal of latent variable SEM for studies in the social sciences is easy to understand. Many, if not most of the, key concepts in the social sciences are not directly observable. The initial phases of

Table 8.1 Objectives encompassed by various types of research

	Objective	Task	Tools (examples)	Tradition
Exploratory	Explore available data in search of a model specification	Data reduction, pattern recognition	Human intuition and pattern recognition; statistical pattern recognition (neural nets, factor analysis)	Interpretive
Confirmatory	Find out how well a model derived from existing theory is confirmed by a particular dataset	Methods that measure how consistent observations are with theory	Statistical hypothesis testing	Descriptive-empirical
Discriminant	Discriminate one model from another by determining which one is better supported by the data	Methods that measure how consistent observations are with one model versus another	Statistical hypothesis testing, pattern recognition, discriminant analysis	Descriptive-empirical
Predictive	Predict the existence and causal direction of potential relationships between candidate factors	Models that predict future observations, even if they seem not to be consistent with historical observations	Econometric forecasting models, neural networks	Positive

Table 8.2 Research traditions

	Approach	Objective	Control over research context	Explicitness of data collection procedures	Example
Interpretive/ qualitative	Synthetic/ holistic	Heuristic/ hypothesis generating	Low, subjective and personal	Low	Specification search, discovery, creation of new theory
Descriptive/ empirical	Analytical/ statistical	Hypothesis testing/theory confirmation	Low, nonintrusive; deals with naturally occurring phenomena	High	Confirmation of existing theory
Positive/ predictive	Analytic/ synthetic/ holistic	Accurate prediction; policy formulation	High if the goal is policy formulation and enactment	High	Schrödinger equation which predicts well despite controversy over what in nature it actually describes

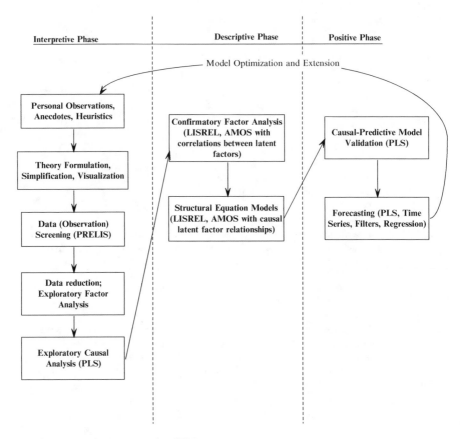

Fig. 8.1 Research programs using SEM

nearly any research project typically involve a modicum of ad hoc pattern recognition. But at a certain point, structure is imposed, and the form that the research takes is determined by the:

1. Data that can be cost effectively acquired
2. Hypotheses worth testing, and where they fit into larger theories
3. Objectives appropriate for the specific research at hand

Tool selection, observation, survey instruments, data sources, statistics, and reporting formats—are all dictated by these three factors (Fig. 8.1).

Possibly the most compelling feature of modern path analysis tools such a PLS path analysis, AMOS, and LISREL) is their ability to tease out network relationships of unobservable but theorized constructs. The choice of latent variable statistical methods arises through a dialectic arising from the need to cost effectively collect objective observations in research disciplines that mainly theorize about subjective and unobservable constructs. Without SEM path analysis tools, hypotheses testing tends to occur indirectly, leaving substantial opportunities for questioning their results and interpretation. Thus SEM path analysis approaches promise the direct testing of hypotheses about unobservables, but at a cost in complexity, and perhaps difficulty in interpreting exactly what the SEM statistical analysis concludes about each hypothesis.

Herman Wold suggested the concept of "plausible causality," a concept that was more completely developed in (Anderson, 1983). Wold's convergence on plausible causality took several turns over its development. Initially after Tukey's (1954) suggestion that path analysis should adopt regression rather than correlation path coefficients, Wold explored both holistic analysis of variance approaches and piecewise path approaches. His main criticism of the analysis of variance approaches was that they failed to address non-normal data, and that most data encountered by researchers is non-normal and often highly multicollinear (Wold, 1980). Since you could not effectively render an opinion on whether the causal links in a model were true or not, one could at best conclude that these links were "causally plausible."

8.5 Hypothesis Testing

Hypothesis tests present a simplified model of the real world that can either be "confirmed" or "rejected" (thus the term "confirmatory analyses") through analysis and summarization of data relevant to the underlying theory. Hypothesis testing draws from deeper philosophical inquiries, best articulated in Popper's (1959) arguments that falsifiability of statements was fundamental to science. Kuhn (1962) further conjectured that scientists work within a conceptual paradigm that influences their role for data, and will go to great length to defend their paradigm against falsification. Lakatos (1970) argued that changing a 'paradigm' is difficult, as it requires an individual scientist to break with peers, but is the duty of scientists. No hypothesis can be unequivocally confirmed through data analysis; instead they are confirmed or rejected with some level of significance α and power $1 - \beta$. The power of a statistical test is the probability that the test will reject a false null hypothesis. Significance α and power $1 - \beta$, along with the distribution of the estimator, will determine the minimum sample size that is required to perform the test (Fig. 8.2).

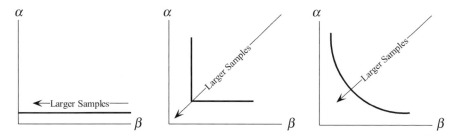

Fig. 8.2 Choice and type I/II (i.e., α and β) error trade-off with (1) fixed significance hypothesis tests; (2) minimax; (3) cost-benefit (e.g., Bayes risk) objective functions, respectively, from *left* to *right* in the figure

Power analysis can either occur before (a priori) or after (*ex post*) data is collected. A priori power analysis is conducted prior to the conducting of research and is typically used to determine an appropriate sample size to achieve adequate power. Post hoc power analysis is conducted after a study has been conducted and uses the obtained sample size and effect size to determine what the power was in the study assuming that the effect size in the sample size is equal to the population effect size. Statistical power depends on:

1. The statistical significance criterion used in the test
2. The size of the difference or the strength of the similarity (that is, the effect size) in the population
3. The sensitivity of the data, or conversely, how much information the dataset has about the particular research question being studied in the hypothesis and theory

Calculating the power requires first specifying the effect size you want to detect. The greater the effect size, the greater the power. Using statistical controls can increase sensitivity, by increasing the reliability of measures (as in psychometric reliability), and by increasing the size of the sample. Increasing sample size is the most commonly used method for increasing statistical power.

Funding agencies, ethics boards, and research review panels frequently request that a researcher perform a power analysis. The argument is that if a study is inadequately powered, there is no point in completing the research. Although there are no formal standards for power, most researchers who assess the power of their tests use $1 - \beta = 0.80$ as a standard for adequacy following Cohen (1977).

8.6   Model Specification and Confirmation

Analysis of data will nearly always involve multiple steps of model specification, data collection, and re-specification. Model confirmation has two sides: (1) supporting the research hypotheses and (2) rejecting competing theories. Difficulties in rejecting competing theories are one reason to favor simpler models (which lead to fewer competitor models).

If model richness and inclusion of unobservable factors are required—which is the case in much of social science research—then analysis of variance and systems of regression methods improve over piecewise estimation in traditional and PLS path analysis because these impose numerous restrictions on data collection (e.g., normality of distributions) and SEM definition (e.g., identification and theory driven modeling). These restrictions mean that the researcher has to work harder for the estimates to be computed, but can make a much stronger case for model validity once the method is coaxed into generating an estimate. In many cases, the restrictions simply cannot be met—this is often true in the case of assuring that the observations are normally distributed. Censored, truncated, or ordinal data will not be normal (though one can test and argue for them being nearly normal). In such cases, weaker arguments may be generated or the researcher may just wait for either a better theory, more data, or both.

It is in addressing such combinatorial choice issues in descriptive research that
the causal-positivist objectives of Wold's plausible causality are able to assist in
model choice. Confirmatory tests of SEM models can be augmented by positive
research that proposes alternative arrangements of latent factors and alternative
causal relationships, and perhaps which adds new latent constructs while eliminat-
ing old. PLS path analysis is suited for such tasks because it does not demand
knowledge of underlying distributions, equation identification, nor large datasets. It
is designed to predict the strength of the model, and the strength of causal relationships between latent factors that can then be used to fine-tune the hypotheses and
the model tests for the next round of descriptive testing.

Wold (1980) suggested that model "confirmation" (i.e., the acceptance of a
hypothesis as to its "truth") occurs under a very broad range of circumstances,
including small datasets and complex models, emphasizing that model predictions
can only be considered *plausible*, rather than confirmed by the data testing. At the
basis of this is the concept that rather than the researchers' a priori hypothesized
model being shown to be the only real model, we instead allow the existence of
alternative models, indicating that this is one plausible model among several others.
How many others? We look at this further in the next section.

8.7 How Many Alternative Models Should You Test?

The stronger the correlations on an SEM path, the more power the method needs to
have to detect an incorrect model. When correlations are low, the researcher may
lack the power to reject the model at hand. Also, the various competing SEM methods tend to overestimate goodness of fit for small samples of less than 200
(Kleinberg, 2000). Similarly, one can have good fit in a misspecified model.
Equivalent models exist for almost all models, and the number of candidate models
grows exponentially with the number of variables. Though systematic examination
of equivalent models is still rare in practice, such examination is increasingly recommended (Scheines, Spirtes, Glymour, Meek, & Richardson, 1998).

To gain a better insight into the alternative candidate models available, consider
that confirmatory testing in descriptive research is basically a process of selecting
one model over its alternatives. From a Neyman-Pearson perspective, it may be
presented as a binary choice between accepting or rejecting a single (null) hypothesis; but in the larger research program, this has to be seen as choosing one hypothesized description of reality (i.e., the model) over many others. How many? This
depends on the complexity of hypotheses tested, and grows combinatorially large as
the statement of the hypothesis grows more complex. Here is an example of the
magnitude of the problem in a path analysis context. Assume that your model has
six (6) constructs (bubbles) (Fig. 8.3).

But you do not a priori know the precise relationships between constructs. In
fact, with six constructs, what you can explore up front is a model with 15 not 5
links. And the number of links you have to explore goes up by the square of the
number of bubbles (Fig. 8.4).

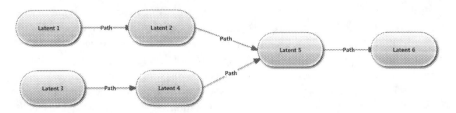

Fig. 8.3 Six constructs and five causal links

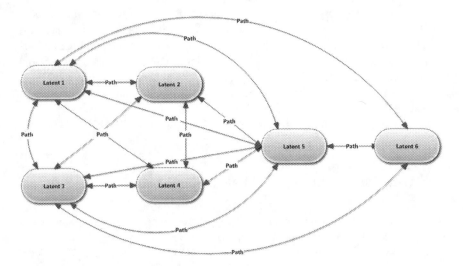

Fig. 8.4 Six constructs and all 15 potential causal relationships

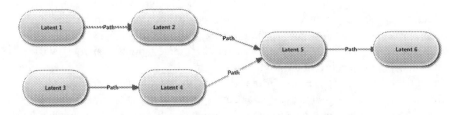

Fig. 8.5 Forward causality on the Latent 1 ~ Latent 2 link

Now for discriminant analysis, you would assume that each one of these links can take a value of 0, +1, or −1: The causal relationship (arrow) either points forward or backward or does not exist (Figs. 8.5, 8.6, and 8.7).

The total number of possibilities in such a discriminant model is $3L$ where L is the number of links between constructs. In the six (6) construct SEM this is $3^{15} = 14,348,907$ distinguishable sets of causal relationships. The number of

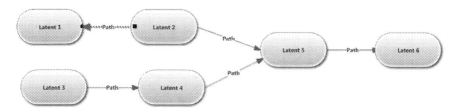

Fig. 8.6 Backward causality on the Latent 1 ~ Latent 2 link

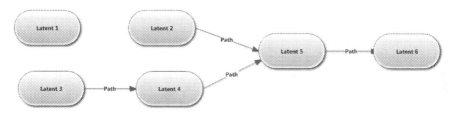

Fig. 8.7 No causality on the Latent 1 ~ Latent 2 link

distinguishable sets of causal relationships rises exponentially in the number of variables. If one assumes, for example, that it takes a minimum of 30 observations to reliably establish the underlying distribution of observations on a single-model parameter, you can see how the sample requirements can grow large quickly.

8.8 Distributional Assumptions

Though the above studies provide some guidance on minimum sample sizes required for use of SEM in hypothesis testing, they do not address challenges of working with non-normal data. This is a particular problem with SEM in the social science, because of the widespread use of Likert scale survey data—5- and 7-bin categorical data.

Jöreskog's maximum likelihood approach generates meaningful goodness-of-fit statistics for the latent SEM if the underlying indicator data is multinormal. PLS path analysis and Jöreskog's other algorithms don't explicitly impose this requirement, but the fit statistics are, unfortunately, difficult to interpret. We know from the studies cited previously that when PLS path analysis models are identified and the indicator data is multinormal, that parameter estimates converge to those of 2SLS, for which goodness-of-fit statistics are well understood. Indeed Goodhue, Lewis, and Thompson (2006, 2012) questioned whether the lack of robustness in the face of non-Gaussian observations should not be reason to abandon PLS path analysis for less lenient approaches such as LISREL.

Wold (1980) emphasized, though, that PLS-PA is robust, and will compute parameter estimates under a very broad range of circumstances, including small datasets and complex models. But beyond this, little is known of the small-sample properties of PLS-PA path estimators, or for that matter how much these properties owe to bootstrapping in their computer implementations. Wold has promoted *predictive-PLS*-PA use in the social sciences, because PLS-PA will yield predictions where other methods cannot, but these predictions can only be considered *plausible*, rather than confirmed by the data testing. If the data is both non-normal and has outliers, the decision to delete values or transform the data is confronted. Transformation of the variable is usually preferred as it typically reduces the number of outliers, and is more likely to produce normality, linearity, and homoscedasticity. In social science work, data is often limited to positive values only, and may be ordinal as well, as is the case for Likert scale responses. Screening will aid in the isolation of data peculiarities and allow the data to be adjusted in advance of further multivariate analysis (Tabachnick & Fidell, 1989).

When the researcher fails to normalize data prior to analysis, then LISREL can search for the best latent variable model fit that will error in several ways. If the maximum likelihood (ML) method is used, standard errors and χ^2 statistics will be incorrect. In theory, weighted least squares procedures using the "correct" weight matrix could produce correct estimates of standard errors and χ^2 statistics, but will still require a substantially larger sample, and assurance that the weights have been properly chosen (which begets further questions).

8.9 Statistical Distributions: Are They Part of the Model or Are They Part of the Data?

Statistical analysis can be dichotomized into (1) analysis that assumes normal or Gaussian distributions and (2) the so-called nonparametric analysis. This may oversimplify, but it does speak to the enormous temptation to "assume" a normal distribution, even where such assumptions are obviously violated.

Where other distributions arise, it is because the processes we are measuring, or the process of measurement itself, have a particular structure. Normal distributions tend to arise when we sum items. Other distributions arise in different situations, for example:

1. Binomial distributions are a result of binary—coin flipping—processes.
2. Categorical distributions from polychotomous—dice throwing—processes; multinomial distributions from classification processes.
3. Zipf-Pareto distributions from measuring rank-frequency.
4. Lognormal distributions arise when we multiply items.
5. Poisson distributions from Poisson processes.

The point to bear in mind is that the distribution of the data depends to some extent on the way the experiment is conducted, choice of factors to measure, and processes that are associated with the reality we are investigating. Before blindly delving ahead and collecting data, it is important to see whether better behaved data (e.g., ones with a normal distribution) can be acquired by altering the design of experiments or data collection.

8.10 Causality

The eighteenth-century Scottish philosopher David Hume observed that "*causes* are a strange sort of knowledge" (Hume, 1758). People talk about *causes* and ostensibly the resulting *effects* as if they were facts about the structure of the universe, to be unraveled and used by scientists for the betterment of mankind. But causes aren't at all factual—rather they are in Hume's words a "lively conception produced by habit." They are the lazy conclusion, the scientific sound bite of modern life. Scientists can discover facts. But cause is not a fact. Cause is a simplification we use to make sense of the facts—it is scientific storytelling.

Causation is a mental shortcut that works well enough most of the time, and places minimal load on mental resources. Belief in causality allows us to safely drive a car (e.g., allowing us to predict that a crash will *cause* an injury) and otherwise make sense of the world. Unfortunately, when it comes to reasoning about complex, interrelated systems, where multiple correlations act simultaneously, cause is not fact. Cause is a mental shortcut that can be dangerously misleading.

In the 1940s, Flemish psychologist Albert Michotte explored the fabrication of causality in his experiments with observers of several short films about a red ball and a blue ball (Michotte, 1963). In the first film, the red ball raced across the screen, touched the blue ball, and then stopped. The blue ball, meanwhile, began moving in the same direction as the red ball. When asked to describe the film, viewers recalled that the red ball hit the blue ball, which *caused* it to move. Michotte called this the launching effect, and found it to be a universal property of visual perception. Although there was nothing about causation in the 2-s film, viewers couldn't help but concoct a story about what had happened—they translated a sequence of stills into a causal story.

Michotte subsequently manipulated the films, asking subjects to describe the changes that took place in the new footage. For example, when he introduced a 1-s break between the movement of the balls, the impression of causality disappeared. The red ball no longer "caused" the blue ball to move. Michotte went on to conduct over 100 similar experiments. For example in one, a small blue ball moved in front of a big red ball—subjects described this as the red ball "chasing" the blue ball. If a big red ball was moving in front of a small blue ball, subjects described the blue ball as "following" the red ball. Chasing and following were alternative story words that had replaced cause and effect.

Michotte drew two conclusions from his experiments. First, our theories about cause and effect are inherently perceptual and subject to the visual shortcuts hardwired into our minds. Michotte saw causal beliefs as similar to color perception—particular objects will automatically be described as "red" just as particular situations will be ascribed causality. Second, causality is a mental oversimplification, more useful in a day when humans regularly made split-second fight or flight decisions.

The dangers of cause and effect simplifications have shown up most prominently—and expensively—in medical diagnosis and drug testing, particularly in the past decade. Pharmaceutical R&D is widely based on targeting causal links in metabolic pathways, and blocking them (an approach pioneered in the nineteenth century by Paul Erlich, which he called the "magic bullet"). In contrast Barabási and Oltvai (2004) provide substantial evidence for perceiving these as metabolic networks rather than pathways, which revisits the arguments that lead from Sewall Wright's path analysis to more holistic approaches to SEM estimation. This matters today, because the R&D required in discovering a new drug candidate is inflation adjusted 100 times what it was in the 1950s. The average cost for approved drug molecule by 2015 is predicted to be $3.5 billion (Lehrer, 2009). Despite this, European regulators have found that 85 % of approved new drugs work poorly or not at all.

The consequences of oversimplification of metabolic pathways using chains of causal links have proved expensive in recent years. Lilly has in the past 2 years discarded two Alzheimer's drugs Semagacestat (which targets amyloid protein metabolism, but failed to do that, and also increased risk of skin cancer) and Dimebon (which targets mitochondria, but fails to slow Alzheimer's disease) after spending billions of dollars in R&D. Pfizer was similarly forced to halt Phase III trials of cholesterol drug Torcetrapib, finding that it actually increased heart failure with a 60 % higher mortality rate. These failures could all be interpreted as the consequences of reasoning on too simple a causal model of disease.

How could these and many other studies have been so wrong? Metabolism is increasingly being understood to be a complex network of chemical interrelationships, where individual compounds may perform multiple services in multiple subsystems. Thinking of these in terms of linear metabolic pathways leads to incorrect models, and failed predictions of drug effectiveness. The simplifications of cause and effect result in misleading, wasteful, and dangerous science that is costing the pharmaceutical industry dearly.

Why is this important structural equation modeling? Because structural equation models almost invariably represent complex networks of inferred behavior that cannot be reduced to binary cause-and-effect relationships. If they could be reduced to simple relationships, we would estimate them as a series of regressions or ANOVAs. In modern society, the elaborate fictions of causality cost us dearly, and are a fundamental challenge in model building.

8.11   The Risks of Received Wisdom: A Case Study

Functional magnetic resonance imaging (fMRI) which measures change in blood flow related to neural activity in the brain and spinal cord has become a core tool in spinal diagnosis in the past two decades. Yet diagnosis has suffered a crisis of scientific method similar to that of drug R&D. Americans spend $90 billion annually treating back pain—roughly the same amount as is spent on cancer. Until the 1970s, the only remedy for back pain was bed rest, and most patients improved. With the advent of MRI in the 1990s, epidurals and surgery were increasingly prescribed, to treat the various "causes" of back pain discovered in fMRI scans. Patient recovery declined.

Enormous amounts of data are generated by fMRI scans, and this data is error prone and difficult to interpret as brain processes are complex and often non-localized. Partial least squares structural models have been widely applied in fMRI imaging to analyze and interpret the mass of data generated in these 15–20-min scans. Unfortunately, such analyses are only as accurate as the data, and are limited by the reliability of the brain model that is used to describe the structural model's latent variables and factors.

The problem of fMRI data accuracy is illustrated in a widely cited article (Bennett, Baird, Miller, & Wolford, 2009) that describes Dartmouth neuroscientist Craig Bennett's fMRI scan of a whole Atlantic salmon (purchased at a local fish market, and which, as Bennett dryly notes, "was not alive at the time of scanning"). While the fish sat in the scanner, Bennett showed it "a series of photographs depicting human individuals in social situations." To maintain the rigor of the protocol the salmon, just like a human test subject, "was asked to determine what emotion the individual in the photo must have been experiencing." If that were all that had occurred, the salmon scanning would simply live on in Dartmouth lore as a "crowning achievement in terms of ridiculous objects to scan." But the fish had a surprise in store. When Bennett got around to analyzing the fMRI data, it looked as if the dead salmon was actually thinking about the pictures it had been shown.

Not only has Bennett's study generated much mirth in a normally tedious field, but it has spawned its own slew of books—from both inside and outside of the field of brain imaging—critical of the statistical methods used to analyze all that data coming out of the fMRI machines (Bennett, Baird, Miller, & Wolford, 2010; Van Rooij & Van Orden, 2011). Law (2010) in particular warns of the implications for use of fMRI in court, as fMRI scans have been promoted as alternatives to inadmissible polygraph evidence.

One lesson should be taken away from these well-documented failures: all of the data collection and analysis in the world will not make up for the failings of a bad model.

8.12 Physics Envy and the Pretentious Model

Social scientists are an insecure lot; and why wouldn't they be? Economics is dismissively referred to as "the dismal science." Physicists like Alan Sokal bait the community with faux articles such as "Transgressing the Boundaries: Towards a Transformative Hermeneutics of Quantum Gravity" (Hair, Sarstedt, Ringle, & Mena, 2012; A. Sokal, 1998; A. D. Sokal, 1996) and (Ringle, Sarstedt, & Straub, 2012) in their book *Higher Superstition* have accused social scientists of practicing the black arts of postmodernist deconstructionism.

Thus it is unsurprising to find that abused social scientists sometimes suffer from *physics envy*—the preoccupation that every process, natural or human, has a basis in something like Newtonian mechanics (despite the fact that modern physics has its quantum mechanics filled with vagaries). There is a tendency (perceived or real) for the so-called softer sciences and liberal arts to try to obtain mathematical expressions of their fundamental concepts, as an attempt to move them closer to harder sciences, particularly physics. Yet the success of physics to mathematicize itself, particularly since Isaac Newton's *Principia Mathematics*, is generally considered as remarkable and often disproportionate compared to other areas of inquiry (Mayr, 2004; Monecke & Leisch, 2012). Propensities towards complex graphs and unnecessary Greek notation are embarrassing symptoms of physics envy.

Science has traditionally bowed to the dictates of Occam's razor (*lex parsimoniae*)—the law of parsimony, economy, or succinctness. It is a principle urging one to select among competing hypotheses that which makes the fewest assumptions and thereby offers the simplest explanation of the effect. Gratuitous complexity contributes to additional fallacies like (1) *complex question* (question presupposing the truth of some assumption buried in that question); (2) *false cause* (one treats as the cause of a thing what is not really the cause of that thing); (3) *apriorism* (refusing to look at any evidence, such as plausible alternative models, that might count against one's claim or assumption); (4) *wishful thinking* (assuming that because one wants something to be true, it is or will be true); and (5) *composition* (reasoning mistakenly from the attributes of a part to the attributes of the whole).

Nowhere are these fallacies more prominently on display than when an SEM path model becomes engorged with constructs—both latent and observed. Not just their sheer number of constructs distinguishes pretentious models, but also their subjectivity. Tension, dissatisfaction, propensity, qualitative overload, pressure, scope, and role conflict are all highly personal and highly subjective value judgments. Any illusion that an overburdened path analysis model is going to reveal deep insights is surely exaggerated.

Modern software makes it easy to throw together complex models without any thought to their validity or usefulness—the computer can always figure out something. Still, the researcher would be well advised to be guided by Occam's Razor, or as Stephen Wolfram paraphrased "it is vain to do with more what can be done with fewer" (Wolfram, 2002).

8.13   Design of Empirical Studies

Freedman (D. A. Freedman, 1987; Merton, 1995) objected to the SEM path analysis failure to distinguish among causal assumptions, statistical implications, and policy claims. Freedman's paper, titled "Statistical models and Shoe Leather" strongly criticized the rigor of quantitative methods in the social sciences. He succinctly articulated the concerns of econometricians towards PLS path analysis and LISREL SEM statistical methods (as well as Wright's original correlation-based path analysis). The main faily of these methods, according to Freedman, was that if the statistical methods *alone* don't insure well-formed hypotheses, proper theory validation, and commensurate data analysis, must be somehow flawed.

The social sciences are at a disadvantage in data collection in comparison with the natural sciences. Historical records like financial statements and surveys of individual behavior tend not only to be subjective, but they are also one-shot, non-replicable measurements. Except in very contrived situations, social scientists find it difficult to set up a controlled lab experiment, and rerun it thousands or millions of times. Furthermore, central constructs in social science theory such as personal utilities are not directly observable. Purely statistical techniques can never solve these problems alone. Effort must be invested in arguing and formulating models and hypotheses. Researchers must dedicate themselves to exploring alternative model specifications, predicting causes and consequences of a well-formed theory, and confirming theory with the data at hand.

It is worth noting that even Bayesian statisticians are the target of variations on this objection. Bayesians presume that there exists prior knowledge about the parameters being estimated (and almost certainly is, if only the expected range of parameter values) and that knowledge can benefit the estimation. Somehow this raises suspicions that Bayesians will "fudge" their priors to obtain a result. Whereas Bayesians explicitly separate out the subjective portion of their estimation, SEM methods and other econometric methods infuse that subjectivity into their hypothesis and underlying models.

SEM estimation places a heavy burden on theory formulation—the model contains both unobserved constructs and causal direction, with complex interactions between unobservables. SEM modeling is highly subjective, and thus the theories underlying SEM must be strong and well argued; alternatives must be proposed; and confirmatory testing needs to be extensive. Like the Bayesians perhaps, social scientists relying on SEM must adhere to a higher standard in their use of a priori subjective information. Subjective model and theory formulation need to be subjected to a consequent greater scrutiny in the exploration and validation of that theory that one would find in the natural sciences or even in econometrics.

In addition, many of the most interesting constructs in the social sciences are not directly observable. As a consequence, we are often forced to conjecture based on observations that we believe are correlated with these unobserved quantities—i.e., observed quantities that somehow "indicate" what is going on with our unobserved "latent" factors.

The quest for knowledge in many of these important yet unobserved (latent) concepts (factors) usually takes one of the three forms—either it is exploratory; theory confirmation; or predictive (Tables 8.1 and 8.2).

Interpretive and positive traditions involve the exploration of our observations for some simpler underlying set of factors. Descriptive theories represented in SEM hypothesize the latent factors underlying observations, and will conduct research to confirm or reject our hypotheses. Finally, we may only be interested in prediction— a much less demanding criterion than confirmation—and indeed may be willing to accept an incorrect model that nonetheless yields accurate predictions. The success of descriptive theory testing in the natural sciences—especially physics—in the twentieth century has tended to bias our research expectations in the social sciences.

Social sciences are disadvantaged by small datasets and inherently unobservable constructs underlying their most important theories. Such circumstances require a heavier investment in the inductive phases of research—interpretive and positive research—than in the theory confirmation of the descriptive phase. Current emphasis on the illusory rigor of descriptive research arises from a legacy of "physics envy" dating back, one could argue, to Adolphe Quetelet in the early nineteenth century. But having developed the tools for more formal theory building, perhaps now is the time for the social sciences to reconsider their emphases, and devote more time to the induction of interpretive and positive research.

8.14   Significance Testing

While many of the measures used in SEM can be assessed for significance, significance testing is less important in SEM than in other multivariate techniques. In other techniques, significance testing is usually conducted to establish that we can be confident that a finding is different from the null hypothesis, or, more broadly, that an effect can be viewed as "real." In SEM the purpose is usually to determine if one model conforms to the data better than an alternative model. It is acknowledged that establishing this does not confirm "reality" as there is always the possibility that an unexamined model may conform to the data even better. More broadly, in SEM the focus is on the strength of conformity of the model with the data, which is a question of association, not significance.

Other reasons why significance is of less importance in SEM are the following:

1. SEM focuses on testing overall models, whereas significance tests are of single effects.
2. SEM requires relatively large samples. Therefore very weak effects may be found significant even for models which have very low conformity to the data.
3. SEM, in its more rigorous form, seeks to validate models with good fit by running them against additional (validation) datasets. Significance statistics are not useful as predictors of the likelihood of successful replication.

8.15   Model Identification

One way is to run a model-fitting program for pretest or fictional data, using your model. Model-fitting programs usually will generate error messages for underidentified models. As a rule of thumb, overidentified models will have degrees of freedom greater than zero in the chi-square goodness-of-fit test. AMOS has a df tool icon to tell easily if degrees of freedom are positive. Note also that all recursive models are identified. Some non-recursive models may also be identified (see extensive discussion by Kline, 1998, ch. 6).

How are degrees of freedom computed? Degrees of freedom equal sample moments minus free parameters. The number of sample moments equals the number of variances plus covariances of indicator variables (for n indicator variables, this equals $n(n+1)/2$). The number of free parameters equals the sum of the number of error variances plus the number of factor (latent variable) variances plus the number of regression coefficients (not counting those constrained to be 1s).

8.16   Negative Error Variance Estimates

When this occurs, your solution may be arbitrary. AMOS will give an error message saying that your solution is not admissible. LISREL will give an error message "Warning: Theta EPS not positive definite." Because the solution is arbitrary, modification indices, t-values, residuals, and other output cannot be computed or are also arbitrary.

There are several reasons why one may get negative variance estimates.

1. This can occur as a result of high multicollinearity. Rule this out first.
2. Negative estimates may indicate Heywood cases (see below).
3. Even though the true value of the variance is positive, the variability in your data may be large enough to produce a negative estimate. The presence of outliers may be a cause of such variability. Having only one or two measurement variables per latent variable can also cause high standard errors of estimate.

For more on causes and handling of negative error variance, see Chen, Bollen, Paxton, Curran, and Kirby (2001).

8.17   Heywood Cases

When the estimated error term for an indicator for a latent variable is negative, this nonsensical value is called a "Heywood case." Estimated variances of zero are also Heywood cases if the zero is the result of a constraint, where without the constraint the variance would be negative. Heywood cases are typically caused by misspecification of the model, presence of outliers in the data, having only two indicators per latent variable, population correlations close to 1 or 0 (causing empirical underidentification), and/or bad starting values in maximum likelihood estimation. It is important that the final model not contain any Heywood cases.

Researchers can resolve a Heywood case by deleting the offending indicator from the model, or by constraining the model by specifying a small positive value for that particular error term. Other strategies include dropping outliers from the data, applying nonlinear transforms to input data if nonlinear relations exist among variables, making sure that there are at least three indicators per latent variable, specifying better starting values (better prior estimates), and gathering data on more cases.

8.18   Empirical Confirmation of Theory

In situations where theory is strong, confirmatory testing and model extension are goals, and appropriate datasets are available, the mainstream statistical approaches such as regression approaches developed at the Cowles Commission and Jöreskog's maximum likelihood SEM approaches are most appropriate. For confirmatory testing, with well-understood fit indices and statistical measures, LISREL-ML is the tool of choice. Jöreskog has provided the PRELIS tools (and AMOS provides similar tools through its underlying SPSS) to filter and transform datasets, so they meet these conditions, without robbing them of explanatory power. Nonetheless, complex models of latent factors require significantly more work and more expense than other methods discussed here (Table 8.3).

Table 8.3 Comparing SEM path analysis methods

	PLS path analysis	Covariance structure methods	Systems of equation regression
Ideal applications	Prediction, specification search	Theory exploration and confirmation	Theory exploration and confirmation, hypothesis testing
Hypothesis testing?	n/a	Likelihood ratio test on observed versus theoretical value of the dispersion matrix	Confidence interval procedures; provides clearest roles for observations
Distributional assumption on indicators	None except that all indicators must have finite variance	Multinormal	Multinormal, but analysis of residuals and transformation allow options for non-normal data
GUI	Yes	Yes	No
i.i.d. residuals?	No	Yes	Yes
Meaning of lines between latent factors	Canonical correlations	Covariances	Regression parameters on latent variables constructed from formative links
Full information?	Limited	Full	Full
Solution process	Iterative search	Iterative search	Closed form algebraic
Solution concept	Least squared error fit on pairs of variables	Maximum likelihood assuming normal distribution	Least squared error fit
Fit measure and accuracy concept	No overall fit statistic	Many	Many
Identifiability	No identification problem	Covariance structure is defined by the block structure of the model, the model may not be identified, and will have to be reparameterized	Rank condition, order condition

References

Anderson, C. A. (1983). The causal structure of situations: The generation of plausible causal attributions as a function of type of event situation. *Journal of Experimental Social Psychology, 19*(2), 185–203.

Barabási, A. L., & Oltvai, Z. N. (2004). Network biology: Understanding the cell's functional organization. *Nature Reviews Genetics, 5*(2), 101–113.

Bennett, C. M., Baird, A. A., Miller, M. B., & Wolford, G. L. (2009). Neural correlates of interspecies perspective taking in the post-mortem Atlantic Salmon: An argument for multiple comparisons correction. *Organization for Human Brain Mapping Abstracts*

Bennett, C. M., Baird, A. A., Miller, M. B., & Wolford, G. L. (2010). Neural correlates of interspecies perspective taking in the post-mortem Atlantic Salmon: An argument for multiple comparisons correction. *Journal of Serendipitous and Unexpected Results, 1*(1), 1–5.

Box, G. E. P., & Draper, N. R. (2007). *Response surfaces, mixtures, and ridge analyses* (Vol. 527). Hoboken, NJ: Wiley-Interscience.

Chen, F., Bollen, K. A., Paxton, P., Curran, P. J., & Kirby, J. B. (2001). Improper solutions in structural equation models causes, consequences, and strategies. *Sociological Methods & Research, 29*(4), 468–508.

Freedman, D. A. (1987). As others see us: A case study in path analysis. *Journal of Educational and Behavioral Statistics, 12*(2), 101–128.

Freedman, D. H. (2010). Lies, damned lies, and medical science. *The Atlantic, 306*(4), 76–84.

Goodhue, D., Lewis, W., & Thompson, R. (2006) PLS, small sample size, and statistical power in MIS research. System Sciences, 2006. HICSS'06. Proceedings of the 39th Annual Hawaii International Conference on. Vol. 8. IEEE.

Goodhue, D., Lewis, W., & Thompson, R. (2007). Research note-statistical power in analyzing interaction effects: Questioning the advantage of PLS with product indicators. *Information Systems Research, 18*(2), 211–227.

Goodhue, D., Lewis, W., & Thompson, R. (2012). Does PLS have advantages for small sample size or non-normal data? *Mis Quarterly, 36*(3), 891–1001.

Gould, S. J. (1977). *Ontogeny and phylogeny*. Cambridge, MA: Harvard University Press.

Hair, J. F., Sarstedt, M., Ringle, C. M., & Mena, J. A. (2012). An assessment of the use of partial least squares structural equation modeling in marketing research. *Journal of the Academy of Marketing Science, 40*(3), 414–433.

Hume, D. (1758). *An enquiry concerning human understanding: A critical edition* (Vol. 3). Oxford: Oxford University Press (reprint 2000).

Ioannidis, J. P. A. (2005). Why most published research findings are false. *PLoS Medicine, 2*(8), e124.

Ioannidis, J. P. A., Haidich, A. B., Pappa, M., Pantazis, N., Kokori, S. I., Tektonidou, M. G., … Lau, J. (2001). Comparison of evidence of treatment effects in randomized and nonrandomized studies. *JAMA, 286*(7), 821–830.

Kleinberg, J. M. (2000). Navigation in a small world. *Nature, 406*(6798), 845.

Kline, R. B. (1998). Software review: Software programs for structural equation modeling: Amos, EQS, and LISREL. *Journal of Psychoeducational Assessment, 16*(4), 343–364.

Law, J. (2010). Cherry-picking memories: fMRI-based lie detection in the US courts

Law, J. R. H. (2011). Cherry-picking memories: Why neuroimaging-based lie detection requires a new framework for the admissibility of scientific evidence under FRE 702 and Daubert. *Yale Journal of Law and Technology, 14*, 1.

Lehrer, J. (2009). *How we decide*. New York, NY: Houghton Mifflin Harcourt (HMH).

Liberati, A., Altman, D. G., Tetzlaff, J., Mulrow, C., Gøtzsche, P. C., Ioannidis, J. P. A., … Moher, D. (2009). The PRISMA statement for reporting systematic reviews and meta-analyses of studies that evaluate health care interventions: Explanation and elaboration. *Annals of Internal Medicine, 151*(4), W-65–W-94.

Mayr, E. (2004). *What makes biology unique?: Considerations on the autonomy of a scientific discipline*. Cambridge, MA: Cambridge University Press.

McCarthy, M. I., Abecasis, G. R., Cardon, L. R., Goldstein, D. B., Little, J., Ioannidis, J. P. A., & Hirschhorn, J. N. (2008). Genome-wide association studies for complex traits: Consensus, uncertainty and challenges. *Nature Reviews Genetics, 9*(5), 356–369.

Merton, R. K. (1957). Priorities in scientific discovery: A chapter in the sociology of science. *American Sociological Review, 22*, 635–659.

Merton, R. K. (1968a). The Matthew effect in science. *Science, 159*(3810), 56.

Merton, R. K. (1968). Science and democratic social structure. *Social Theory and Social Structure*, 604–615.

Merton, R. K. (Ed.). (1968). *Social theory and social structure*. New York, NY: Simon and Schuster.

Merton, R. K. (1988). The Matthew effect in science, II: Cumulative advantage and the symbolism of intellectual property. *Isis, 79*, 606–623.

Merton, R. K. (1995). The Thomas theorem and the Matthew effect. *Social Forces, 74*, 379–422.

Michotte, A. (1963). *The perception of causality*. London, UK: Methuen.

Monecke, A., & Leisch, F. (2012). semPLS: Structural equation modeling using partial least squares. *Journal of Statistical Software, 48*(3), 1–32.

Ringle, C. M., Sarstedt, M., & Straub, D. W. (2012). Editor's comments: A critical look at the use of PLS-SEM in MIS quarterly. *MIS Quarterly, 36*(1), iii–xiv.

Scheines, R., Spirtes, P., Glymour, C., Meek, C., & Richardson, T. (1998). The TETRAD project: Constraint based aids to causal model specification. *Multivariate Behavioral Research, 33*(1), 65–117.

Sokal, A. D. (1996). Transgressing the boundaries: Toward a transformative hermeneutics of quantum gravity. *Social Text, 46/47*, 217–252.

Sokal, A. (1998). What the social text affair does and does not prove. *Critical Quarterly, 40*(2), 3–18.

Stigler, S. M. (1980). Stigler's law of eponymy*. *Transactions of the New York Academy of Sciences, 39*(1 Series II), 147–157.

Tabachnick, B. G., & Fidell, L. S. (1989). *Using multivariate statistics* (2nd ed.). New York, NY: HarperCollins.

Tukey, J. W. (1954). Causation, regression, and path analysis. In O. Kempthorne et al. (Eds.), *Statistics and mathematics in biology* (pp. 35–66). Ames, IA: The Iowa State College Press.

Van Rooij, M., & Van Orden, G. (2011). It's about space, it's about time, neuroeconomics and the brain sublime. *The Journal of Economic Perspectives, 25*(4), 31–55.

Wold, H. (1980). *Model construction and evaluation when theoretical knowledge is scarce*. New York, NY: Academic.

Wolfram, S. (2002). *A new kind of science* (Vol. 5). Champaign, IL: Wolfram media.

Chapter 9
From Paths to Networks: The Evolving Science of Networks

Path models were always a kludge; a hodgepodge of available technologies cobbled together, as best as possible, to make sense of naturally occurring networks. Scientists in the past simply did not possess the analytical power to map more than a few links at a time. PLS-PA, LISREL, and systems of regressions were designed for calculation on paper and with adding machines; they were disappointingly inadequate, but the best we had at the time. Statistical power has always lagged the size and complexity of the networks under analysis, and as a result generated unreliable, simplistic, and inapplicable results. This is doubly unfortunate when we consider how important network models have reigned throughout mankind's history. For example:

1. The Romans were obsessed with water networks of aqueducts, plumbing, and hydraulic networks. Romans visited public baths daily and upper class homes were centered on an interior pond. Hydraulic networks defined the medicine (bodily humors), science (hydraulics), and business systems (roads, canals, and pipes) of the Romans.
2. Hydraulic empires—Egypt, Somalia, the Ajuran Empire, Sri Lanka, Mesopotamia, China, Aztec, Maya, and Indus Valley civilizations—were government structures of the largest ancient civilizations. These exercised power through exclusive control over access to water. Agricultural wealth, which accumulates around rivers and their arteries, created opportunities for hydraulic despotism through flood control and irrigation. Imperial bureaucracies required deep knowledge of hydraulic networks to rule and thrive (Mitchell, 1973; Pryor, 1980; Wittfogel, 1957).
3. Hereditary and fealty networks defined the governments in the medieval world, and even into some twenty-first-century regions. In feudal societies, politics and war were won or lost based on control of hereditary and fealty networks.
4. Genetic, metabolomic, proteomic, and epidemiological webs throughout the twentieth century finally gave medicine a firm empirical and scientific foundation, allowing them to build on discoveries in chemistry, physics, and mechanics.

© Springer International Publishing Switzerland 2015
J.C. Westland, *Structural Equation Models*, Studies in Systems, Decision and Control 22, DOI 10.1007/978-3-319-16507-3_9

5. In the twenty-first century, networks have so far set the agenda for new paradigms in business, biology, sociology, computer science, finance, marketing, and
   many other fields.

Each age has evolved its signature paradigms for linking the myriad networks
structuring their worlds to the empirical reality deciding survival or extinction. The
twenty-first century is differentiated by our acquisition of powerful network analysis tools using superfast computers with limitless storage directed by sophisticated
algorithms. This chapter surveys the rapid evolution of computerized network analytics that hint at a deeper science that is only currently evolving.

9.1 Genetic Pathways Revisited

Classical genetic mapping through pedigree analysis and breeding experiments
could be used to determine sequence features within a genome, though the methods
were time consuming with inherently low resolution. This was the world of Gregor
Mendel and Sewall Wright; it was the world of nineteenth-century dog breeders. In
contrast, modern molecular gene mapping techniques are usually referred to as
physical mapping—they use data from gene chips that measure which genes are
active, or expressed, in a cell. Network analysis is using data from molecular gene
maps to provide us with a much more detailed and nuanced picture of life in all of
its complexity.

Amid thousands of studies using such chips, many compared the gene activity
patterns in diseased tissue with that of healthy tissue. The number of genes associated with diseases is expanding rapidly because of so-called whole-genome association studies. In these studies, gene chips are used to look for differences between the
genomes of people with a disease and those without. Much of the raw data from
such studies are deposited in databases, allowing researchers to now gather data on
gene activity for scores of diseases and perform statistical analyses to map diseases
based on similarities in their patterns of gene activity.

Physicist Albert-László Barabási has been pushing the bounds of empirical analysis of networks for the past two decades, applying network theory to problems in
the social sciences, commerce, physics, mathematics, and computer science. He has
studied the growth and preferential attachment mechanisms responsible for the
scale-free structure of the World Wide Web. But his most exciting studies have been
in genetic networks, harkening back to the origins of network path analysis by
Gregor Mendel and Sewall Wright. Barabási and his colleagues obtained lists of
disorders, disease genes, and their associations from the *Online Mendelian
Inheritance in Man* database, compiling information on 1,286 disorders and 1,777
disease genes (A. L. Barabási, 2007; A. L. Barabási, Gulbahce, & Loscalzo, 2011;
A. L. Barabási & Oltvai, 2004; Goh et al., 2007). Starting from a bipartite "diseasome" graph, they generated two network projections: (1) a human disease network
that connected disorders to each other that share a common disease gene and (2) a
disease gene network that connected genes together that are associated with a common

disorder. Diseases were represented by circles, or nodes, and linked to other diseases by lines that represent genes they have in common.

The human disease and gene network reveal the role of peripheral proteins for diseases caused by a variety of genetic mutations. Some diseases, such as Tay-Sachs, result from different mutations in a single gene, whereas other diseases, such as Zellweger syndrome, are caused by a mutation in any one of multiple genes. Generally, cancers were caused by somatic genetic mutations in essential or house-keeping genes. However, most inherited disease genes localized to the functional periphery of the network, with mutations preferentially in nonessential genes.

That Barabási's research is changing the field of disease classification is known. Seemingly dissimilar diseases are being lumped together, and what were thought to be single diseases are being split into separate ailments. For example, two tumors that arise in the same part of the body and look the same on a pathologist's slide might be quite different in terms of what is occurring at the gene and protein level. Certain breast cancers are already being treated differently from others because of genetic markers like estrogen receptor and Her2, and also more complicated patterns of genetic activity. Researchers can profiles drugs by the genes they activate as a way to find new uses for existing drugs. Such research can fundamentally alter and improve our understanding of the causes of disease and of the functions of particular genes. For instance, two genes have recently been found to influence the risk of both diabetes and prostate cancer.

But Barabási's network analysis advances medicine at a much more organic level. Providing a consistent way to classify diseases is also essential for tracking public health and detecting epidemics. The World Health Organization takes pains to periodically revise its International Classification of Diseases, which is used, among other ways, to tally the causes of death throughout the world. The classification is also the basis of the ICD-9 codes used for medical billing in the USA. The first international classification, in the 1850s, had about 140 categories of disease; the tenth edition, in 1993, had 12,000 categories. The increase stems mainly from better knowledge and diagnostic techniques that allow diseases to be distinguished from one another. For most of human history, diseases were named and classified by symptoms, which was all people could observe.

Up to the eighteenth century, Aristotle and Galen were the primary references for medical knowledge. Linnaeus developed a symptom-based taxonomy of disease with 11 classes—painful disease, motor diseases, blemishes, and so on—that were further broken down into orders and species. Doctors who emphasized empirical observation, such as the surgeon John Hunter, were too often ignored by the medical establishment. By the nineteenth century, surgery had advanced to the point where diseases began to be classified by their anatomic or physiological features. The stethoscope let doctors realize that what had been thought of as 17 conditions—like coughing up blood and shortness of breath—could all be different symptoms of the same disease, tuberculosis.

Genetic networks allow the study of diseases at a finer level than even physiological tests. Genes are the instructions for the production of proteins, which interact in complex ways to carry out functions in the body. Disruptions in these molecular pathways can cause disease. Diseases have been subdivided by the type of mutation. Hemophilia was divided into hemophilia A and B, caused by mutations in different genes for different clotting factors. And what was once considered a mild form of hemophilia was later identified as a variant of a different clotting disorder, von Willebrand disease, caused by mutations in a different gene and requiring a different clotting factor as treatment. In contrast, two rare syndromes with different symptoms might represent a continuum of one disease. One syndrome, Meckel-Gruber, is tied to neural defects and death in babies. The other, Bardet-Biedl, is marked by vision loss, obesity, diabetes, and extra fingers and toes.

9.2   The Evolution of SEM Research Questions

Sewall Wright's research was focused on teasing out an empirical understanding of Mendelian inheritance of biological traits long before there was any understanding of the basis underlying biological taxonomies. His objective was the *creation of valid models of inheritance* from raw data, without the benefit of preexisting theories. Observations could be replicated theoretically without end, simply by breeding another generation.

Applications of computer-intensive canonical correlation to Wright's path analysis by Herman Wold and his student Karl Jöreskog were designed to *prove or disprove theories about the structural relationships between unobservable quantities in social sciences*. Applications started with economics, but found greater usefulness in measuring unobservable model constructs such as intelligence, trust, and value in psychology, sociology, and consumer sentiment.

Systems of regression equation approaches pioneered by Tjalling Koopmans were designed to *prove or disprove theories about the structural relationships between economic measurements*. Because regression fits actual observations rather than abstract concepts, they are able to provide a wealth of goodness-of-fit information to assess the quality of the theoretical models tested. Wold and Jöreskog's methods provide goodness-of-fit information that in comparison is sparse, unreliable, and difficult to interpret.

Social network analysis extends graph theory into empirical studies. It takes observations (e.g., Wright's genetic traits) and classifies them as '"nodes" (also called "vertices"). It infers relationships, called "edges" or "links," between these nodes from real-world observations. Social networks reflect social relationships in terms of individuals (nodes) and relationships (links) between the individuals. Examples of links are contracts, acquaintances, kinship, employment, and romantic relationships. A social network may be undirected, meaning that there is no distinction between the two nodes associated with each link, or its links may be directed from one node to another. In Wold's path analysis links are inherently undirected, but Wold invites researchers to "plausibly infer" that links between unobserved variables have a direction (Basmann, 1963). Jöreskog's approach distinguishes between directed (path) links and undirected (confirmatory factor analysis) links between latent variables. Paths between observations and latent variables are always assumed to be directed. But in all of these cases, link direction derives from the researcher's a priori model building, not from empirical tests and data analysis. Further adding to complexity of analysis, when modeling relations between two different classes of things, bipartite graphs will arise naturally. For instance, a graph of scientists and journals they have published in, with an edge between the scientist and the journal if the player has published in that journal, is a natural example of an affiliation network, a type of bipartite graph used in social network analysis.

Research on social networks as a discipline began in the mid-nineteenth century with the work of Karl Marx, Max Weber, and David Émile Durkheim (Calhoun, 2007). Mathematical approaches date from the mid-1950s with studies by Manfred Kochen, an Austrian who had been involved in urban design, and the interconnectedness and "social capital" of human networks, and his colleague Ithiel de Sola Pool, a researcher on technology and society. Kochen and de Sola Pool's manuscript, Contacts and Influences (De Sola Pool & Kochen, 1979), was conceived while both were working at the University of Paris in the early 1950s, during a time when psychologist Stanley Milgram visited and collaborated in their research. Michael Gurevich (1961) contributed empirical studies of the structure of social networks in his 1961 MIT dissertation under de Sola Pool which became part of their unpublished manuscript circulated among academics for over 20 years before publication in 1978. It formally articulated the mechanics of social networks, and explored the mathematical consequences of these, including the degree of connectedness. The manuscript left many significant questions about networks unresolved, and one of these was the number of degrees of separation in actual social networks.

Milgram continued Gurevich's experiments in acquaintanceship networks at Harvard University on his return from Paris. His results were reported in "The Small World Problem" (Milgram, 1967) in the popular science journal *Psychology Today* with a more rigorous version of the paper appearing in *Sociometry* 2 years later (Travers & Milgram, 1969). The *Psychology Today* article generated enormous publicity for the experiments, which are well known today, long after much of the formative work has been forgotten. Milgram showed that people in the USA seemed to be connected by approximately three acquaintance links, on average.

Kochen and de Solla Poole subsequently constructed Monte Carlo simulations based on Milgram's and Gurevich's data which recognized that both weak and strong acquaintance links are needed to model social structure. Kochen worked at IBM at the time, and the simulations, carried out on the relatively limited computers of the 1970s, were nonetheless able to predict that a more realistic three degrees of separation existed across the US population. Their article "Contacts and Influences" (De Sola Pool & Kochen, 1979) became the lead article in the inaugural issue of the journal *Social Networks* and was widely read. It concluded that in a US-sized population without social structure, "it is practically certain that any two individuals can contact one another by means of at most two intermediaries. In a socially structured population it is less likely but still seems probable. And perhaps for the whole world's population, probably only one more bridging individual should be needed." Their peers extrapolated these results to the well-known "six degrees of separation" for global population.

As time progressed, Kochen and de Solla Poole's so-called small world networks (Watts & Strogatz, 1998) were joined by "random" and "scale-free" social network topologies (A. L. Barabási, Dezső, Ravasz, Yook, & Oltvai, 2003) which attempted to place empirical footings under a diverse set of social network topologies. Much of this work takes characteristics that have been studied in graph models, and matches them to particular empirical statistics from real-world networks. They are an important tool for research in sociology, political science, anthropology, biology, communications, finance, economics, bibliometrics, psychology, linguistics, and marketing.

9.3   The New Language of Social Networks

The language of graph theory is rich with descriptors for network properties. Most of these can be applied to social networks such as those that have been the focus of this book. I present here some of the most useful concepts germane to applications covered in this book. The interested researcher may follow up with a more extensive text of graph modeling to gain a more extensive understanding of the vocabulary of networks. Concepts important for path analysis and social network modeling fall into three categories: (1) visualization models that choose how to best present network information for human consumption; (2) link qualifiers and metrics that describe magnitude and qualitative features of relationships between nodes; and (3) topological statistics that summarize more fundamental geometric properties of the network.

9.3.1   Visualization

Visualization for social networks has become popular with the development of software for mapping networks, particularly in 2 dimensions, for display. Dimensions may be added by dynamically adjusting parameter values on a time-lapse video, and by categorizing nodes and links with colors, shapes, and legends. Visualization offers a powerful tool for publicizing data and research, but leaves open many opportunities for visual miscues, aberrations, and illusions unchecked by more rigorous analytical procedures. Particular visualization approaches tend to be chosen for their artistic appeal than for statistical veracity. Force-directed graph drawing algorithms provide one such popular approach for visualizing social networks. They position the nodes so that links are of equal length with few crossings, and then assign springlike forces to the links to place nodes at points of minimum "energy" (Kobourov, 2012). Fruchterman-Reingold (Fruchterman & Reingold, 1991), Force Atlas 1 and 2, and other algorithms are used by software such as Graphviz or Gephi to produce attractive renderings of moderately large datasets. Larger datasets are more difficult to visualize in their entirety, but can be dynamically viewed, much like topological maps, through topological zooming algorithms and constant information density displays.

9.3.2 Link Qualifiers and Metrics

In social network analysis, the vocabulary describing links is much richer than just the terms "directed" and "undirected" links might convey. Links on average may be measured according to:

1. Transitivity: An individual's assumption that his or her friends are also friends (Flynn, Reagans, & Guillory, 2010)
2. Proximity: A tendency for individuals to have more ties with geographically close others (Kadushin, 2012)
3. Homophily: Average similarity of links in terms of gender, race, age, occupation, educational achievement, status, values, and so forth (McPherson, Smith-Lovin, & Cook, 2001)
4. Multiplexity: The number of relationships represented in a single link (Podolny & Baron, 1997)
5. Reciprocity: An extent to which individuals reciprocate (Kadushin, 2012)
6. Strength: Combination of time, emotional intensity, intimacy, and reciprocity (Granovetter, 1973)

Not all of these link properties are necessarily quantifiable, or at least not with any great accuracy. They may simply reflect qualitative social artifacts that researchers would like to better understand.

9.3.3 Topological Statistics

Networks possess geometric properties that are invariant under change of shape or size of networks. The following are some of the most commonly encountered metrics and concepts describing network topologies, along with examples.

1. Connectedness measures: Four concepts describe topological qualities of subsets of nodes and links within a network under study:

    (a) Clique: A completely connected subnetwork, where all nodes are connected to every other node. These networks are symmetric in that all nodes have in-links and out-links from all others.
    (b) Weakly connected component: A collection of nodes in which there exists a path from any node to any other, ignoring directionality of the edges.
    (c) Strongly connected component: A collection of nodes in which there exists a directed path from any node to any other.
    (d) Giant component: A single connected component that contains the majority of the nodes in the network.

2. Centrality: A group of metrics that aim to quantify the "importance" or "influence" of a particular node or set of nodes within a network (Opsahl, Agneessens, & Skvoretz, 2010). Centrality ranks the importance of the various nodes in a

network, with different definitions of "importance" yielding different measures of centrality. If many nodes are connected through one particular node, say a very popular social icon, then that node has high "betweenness centrality." In contrast "eigenvalue centrality" is like a Google PageRank metric that ranks a node as important if many other highly important nodes link to it. The number of such centrality measures is potentially limitless. Such measures are useful in identifying the most central nodes, but are meaningless for other nodes.

3. Distance: The minimum number of ties required to connect two particular actors, as popularized by Stanley Milgram's small world experiment and the idea of "six degrees of separation" (De Sola Pool & Kochen, 1979; Kochen, 1989).

4. "Cliques" are groups where every individual is directly tied to every other individual, "social circles" if there is less stringency of direct contact, which is imprecise, or as structurally cohesive blocks if precision is wanted (White, Owen-Smith, Moody, & Powell, 2004)

5. Clustering: A measure of the likelihood that two associates of a node are associates. A higher clustering coefficient indicates a greater "cliquishness" (White et al., 2004).

6. Cohesion: The degree to which nodes are connected directly to each other by cohesive bonds. Structural cohesion refers to the minimum number of members who, if removed from a group, would disconnect the group (White et al., 2004).

7. Average degree: The degree k of a node is the number of links connected to it. Closely related to the density of a network is the average degree, $k = \frac{2L}{N}$ where L is the number of links and N the number of nodes.

8. Average path length: Is calculated by finding the shortest path between all pairs of nodes, adding them up, and then dividing by the total number of pairs, and shows the average number of steps it takes to get from one member of the network to another. Six degrees of separation describes the average path length between all individuals in the world.

9. Diameter of a network: The longest of all the calculated shortest paths in a network, reflecting the linear size of a network.

10. Density: Ratio of the number of actual links to the number of possible links, a concept closely related to the clustering coefficient.

9.3.4   Network Archetypes

Three fundamental network archetypes are often claimed to describe the known topologies of social networks: (1) random; (2) small-world; and (3) scale-free networks. But this taxonomy may not provide clear boundaries for real-world networks, and there is a high probability that a more precise taxonomy will ultimately prevail. Nonetheless, knowledge of the characteristics of these three topologies is useful in understanding the emerging science of network analytics.

Random networks connect their nodes with links distributed with equal probabilities (Erdős, 1959; Erdős & Rényi, 1960, 1961). Such networks have well-defined properties, and may be used to benchmark other models, especially empirically derived models.

Small-world networks can be represented by a graph in which most nodes are not neighbors of one another, but most nodes can be reached from every other by a small number of hops or steps. Specifically, a small-world network is defined to be a network where the typical distance (number of links in the path) between two randomly chosen nodes grows proportionally to the logarithm of the number of nodes in the network. Small-world networks have a very high clustering coefficient along with high average path length. Each change in linkages is likely to create a shortcut between highly connected clusters (M. E. Newman, 2001; M. Newman, Barabási, & Watts, 2006; M. E. Newman, Moore, & Watts, 2000; M. E. Newman, Watts, & Strogatz, 2002; Watts, 2004; Watts & Strogatz, 1998). Many real-world networks seem to have small-world properties: Facebook's networks, the connectivity of the Internet, wikis such as Wikipedia, and gene networks all exhibit small-world network characteristics. The concept of "six degrees of separation" originally involved small-world network models, but it is also possible in scale-free models to demonstrate that Dunbar's number is the cause of the phenomenon known as the "six degrees of separation" (Dunbar, 1993, 1995; Hernando, Villuendas, Vesperinas, Abad, & Plastino, 2010).

Scale-free networks have a degree distribution that asymptotically follows a power law; the fraction $P(k)$ of nodes in the network having k connections tends to $P(k) \sim k^{-\gamma}$ with $2 < \gamma < 3$. Ideally, this may be considered a random network with a degree distribution following the scale-free ideal gas density distribution. These networks are able to reproduce city-size distributions and electoral results by unraveling the size distribution of social groups with information theory on complex networks when a competitive cluster growth process is applied to the network (Hernando et al., 2010; Moreira, Paula, Filho, Raimundo, & Andrade, 2006). The attribution of "scale-free" to networks has often been used carelessly, with many of the claims being refuted by later research (Clauset, Shalizi, & Newman, 2009). Preferential attachment (A. L. Barabási et al., 2003), the fitness model (Caldarelli, Capocci, De Los Rios, & Muñoz, 2002), and many other mathematical models have attempted to capture the structure of empirical scale-free networks.

9.4   Cross-Pollination: Biological Networks and Banking Ecosystems

Advances in biostatistics and network analysis have encouraged researchers to apply these methods in other, only distantly related fields. For example, zoologist Robert M. May conducted several studies into the stability of the UK banking ecosystems using network models. Haldane and May (2011, May and Arinaminpathy 2010, and May, Levin, and Sugihara 2008) asked how shocks to one or a few banks (at least initially) might propagate through the banking system via loan defaults, Fire Sales, liquidity hoarding, and general "loss of confidence." May modeled UK banks—what he called "nodes in a model financial ecosystem" governed by Basel I and II capital requirements—in a quasi-ecological network. Parameters were set from published bank portfolio holdings, regulatory parameters, and general knowledge of interbank borrowing and lending levels and patterns.

Borrowing methods from plant-pollinator (Bascompte & Stouffer, 2009) and nested hierarchies in food webs (Sugihara & Ye, 2009), May developed a schematic model for a "node" in the interbank network, using them to search for regions of instability. May's model, with parameters set by pre-2008 crash transaction levels, correctly predicted the character and extent of banking failures, and the results of modeling have motivated more recent UK treasury policy. Andrew Haldane (2009) of the Bank of England remarked that May's simulation suggested "One simple means of altering the rules of the asymmetric game between banks and the state is to place heavier restrictions on leverage." He also noted that large capital reserves allow greater robustness of both individual banks and of the system as a whole, and that these should be relatively larger in boom times, when the temptation to take greater risks seems prevalent. Additionally, bigger banks should hold their ratio of capital reserves to total assets at least as high as smaller banks. In practice the contrary is observed.

Perhaps most revealing in May's models are conclusions about what makes for a robust industrial organization of banking. In ecosystems, "modular organization" is often seen, and promotes systemic robustness; this was true of May's models as well. In contrast, the past three decades have witnessed a massive global consolidation and scaling of banking operations, combined with linking of banks through consistently high volumes of interbank loans and transfers. The current banking ecosystem is extremely sensitive to "shocks" to the system.

9.5   The Future

The century-long journey, from dog breeders to the "diseasome" continues to evolve. Sewall Wright developed his method of path analysis to provide a statistical underpinning to the emerging network sciences of mathematical genetics and population studies. These studies were central to the exponential growth in farming productivity—both crops and animals—in the first half of the twentieth century. Griliches (1957) provided some of the earliest statistical support—in his studies of the diffusion of hybrid corn—to the possibility of exponential year-on-year productivity increases brought implicated in Wrights seminal work in the science of genetics and breeding.

Problems in genetics continue to inspire advances in path analysis. Barabási has over the past decade shown how tools borrowed from physics, graph theory, and computer science can be applied to model the exceptionally large and complex causal pathways of protein interactions and metabolism in living organisms. Barabási's research program perceives human metabolism and its genetic foundations in terms of a complexity pyramid of nested causal path models. Lord May has used other network models to gain insight into social and economic problems. And many other researchers continue to expand our understanding of the complexities of social and natural science in this expanding research frontier of network path models. This would have pleased and excited the twentieth century's pioneers in social network analysis tools—from Sewall Wright's genetics; Alfred Cowles' market analyses; Tjalling Koopmans and Lawrence Klien's development of algebraic methods; Herman Wold, Karl Jöreskog, and Trygve Haavelmo's exploration into networks of unobservables; Stanley Milgram, Manfred Kochen, Ithiel de Sola Pool, Mark Granovetter, Duncan Watts and Steven Strogatz' exploration of social and acquaintance networks; and the many others who helped evolve the modern foundations for our understanding of networks of social activities and relationships.

References

Barabási, A. L. (2007). Network medicine—From obesity to the "diseasome". *New England Journal of Medicine, 357*(4), 404–407.

Barabási, A.-L., Dezső, Z., Ravasz, E., Yook, S.-H., & Oltvai, Z. (2003). *Scale-free and hierarchical structures in complex networks*. Paper presented at the Modeling of Complex Systems, Seventh Granada Lectures.

Barabási, A. L., Gulbahce, N., & Loscalzo, J. (2011). Network medicine: A network-based approach to human disease. *Nature Reviews Genetics, 12*(1), 56–68.

Barabási, A. L., & Oltvai, Z. N. (2004). Network biology: Understanding the cell's functional organization. *Nature Reviews Genetics, 5*(2), 101–113.

Bascompte, J., & Stouffer, D. B. (2009). The assembly and disassembly of ecological networks. *Philosophical Transactions of the Royal Society, B: Biological Sciences, 364*(1524), 1781–1787.

Basmann, R. L. (1963). The causal interpretation of non-triangular systems of economic relations. *Econometrica, 31*, 439–448.

Caldarelli, G., Capocci, A., De Los Rios, P., & Muñoz, M. A. (2002). Scale-free networks from varying vertex intrinsic fitness. *Physical Review Letters, 89*(25), 258702.

Calhoun, C. J. (2007). *Classical sociological theory*. Malden, MA: Blackwell.

Clauset, A., Shalizi, C. R., & Newman, M. E. J. (2009). Power-law distributions in empirical data. *SIAM Review, 51*(4), 661–703.

De Sola Pool, I., & Kochen, M. (1979). Contacts and influence. *Social Networks, 1*(1), 5–51.

Dunbar, R. I. M. (1993). Coevolution of neocortical size, group size and language in humans. *Behavioral and Brain Sciences, 16*(04), 681–694.

Dunbar, R. I. M. (1995). Neocortex size and group size in primates: A test of the hypothesis. *Journal of Human Evolution, 28*(3), 287–296.

Erdős, P. (1959). {On random graphs, I}. *Publicationes Mathematicae, 6*, 290–297.

Erdős, P., & Rényi, A. (1960). On the evolution of random graphs. *Publication of the Mathematical Institute of the Hungarian Academy of Sciences, 5*, 17–61.

Erdős, P., & Rényi, A. (1961). On the strength of connectedness of a random graph. *Acta Mathematica Hungarica, 12*(1), 261–267.

Flynn, F. J., Reagans, R. E., & Guillory, L. (2010). Do you two know each other? Transitivity, homophily, and the need for (network) closure. *Journal of Personality and Social Psychology, 99*(5), 855.

Fruchterman, T. M. J., & Reingold, E. M. (1991). Graph drawing by force-directed placement. *Software: Practice and experience, 21*(11), 1129–1164.

Goh, K. I., Cusick, M. E., Valle, D., Childs, B., Vidal, M., & Barabási, A. L. (2007). The human disease network. *Proceedings of the National Academy of Sciences, 104*(21), 8685.

Granovetter, M. S. (1973). The strength of weak ties. *American Journal of Sociology, 78*, 1360–1380.

Griliches, Z. (1957). Hybrid corn: An exploration in the economics of technological change. *Econometrica, 25*, 501–522.

Gurevich, M. (1961). *The social structure of acquaintanceship networks*. Cambridge, MA: MIT Press.

Haldane, A. G. (2009). *Rethinking the financial network*. Speech delivered at the Financial Student Association, Amsterdam, April.

Haldane, A. G., & May, R. M. (2011). Systemic risk in banking ecosystems. *Nature, 469*(7330), 351–355.

Hernando, A., Villuendas, D., Vesperinas, C., Abad, M., & Plastino, A. (2010). Unravelling the size distribution of social groups with information theory in complex networks. *The European Physical Journal B-Condensed Matter and Complex Systems, 76*(1), 87–97.

Kadushin, C. (2012). *Understanding social networks: Theories, concepts, and findings*. New York, NY: Oxford University Press.

Kobourov, S. G. (2012). Spring embedders and force directed graph drawing algorithms. arXiv preprint *arXiv:*1201.3011.

Kochen, M. (1989). *The small world*. Norwood, NJ: Ablex.

May, R. M., & Arinaminpathy, N. (2010). Systemic risk: The dynamics of model banking systems. *Journal of the Royal Society Interface, 7*(46), 823–838.

May, R. M., Levin, S. A., & Sugihara, G. (2008). Complex systems: Ecology for bankers. *Nature, 451*(7181), 893–895.

McPherson, M., Smith-Lovin, L., & Cook, J. M. (2001). Birds of a feather: Homophily in social networks. *Annual Review of Sociology, 27*, 415–444.

Milgram, S. (1967). The small world problem. *Psychology Today, 2*(1), 60–67.

Mitchell, W. P. (1973). The hydraulic hypothesis: A reappraisal. *Current Anthropology, 14*, 532–534.

Moreira, A. A., Paula, D. R., Filho, C., Raimundo, N., & Andrade, J. S., Jr. (2006). Competitive cluster growth in complex networks. *Physical Review E, 73*(6), 065101.

Newman, M. E. J. (2001). Scientific collaboration networks. I. Network construction and fundamental results. *Physical Review E, 64*(1), 016131.

Newman, M., Barabási, A.-L., & Watts, D. J. (2006). *The structure and dynamics of networks*. Princeton, NJ: Princeton University Press.

Newman, M. E. J., Moore, C., & Watts, D. J. (2000). Mean-field solution of the small-world network model. *Physical Review Letters, 84*(14), 3201.

Newman, M. E. J., Watts, D. J., & Strogatz, S. H. (2002). Random graph models of social networks. *Proceedings of the National Academy of Sciences, 99*(Suppl 1), 2566–2572.

Opsahl, T., Agneessens, F., & Skvoretz, J. (2010). Node centrality in weighted networks: Generalizing degree and shortest paths. *Social Networks, 32*(3), 245–251.

Podolny, J. M., & Baron, J. N. (1997). Resources and relationships: Social networks and mobility in the workplace. *American Sociological Review, 62*, 673–693.

Pryor, F. L. (1980). The Asian mode of production as an economic system. *Journal of Comparative Economics, 4*(4), 420–442.

Sugihara, G., & Ye, H. (2009). Complex systems: Cooperative network dynamics. *Nature, 458*(7241), 979–980.

Travers, J., & Milgram, S. (1969). An experimental study of the small world problem. *Sociometry, 32*, 425–443.

Watts, D. J. (2004). The "new" science of networks. *Annual Review of Sociology, 30*, 243–270.

Watts, D. J., & Strogatz, S. H. (1998). Collective dynamics of 'small-world' networks. *Nature, 393*(6684), 440–442.

White, D. R., Owen-Smith, J., Moody, J., & Powell, W. W. (2004). Networks, fields and organizations: Micro-dynamics, scale and cohesive embeddings. *Computational and Mathematical Organization Theory, 10*(1), 95–117.

Wittfogel, K. A. (1957). Oriental despotism. *New Haven, 2*, 251–269.

Index

© Springer International Publishing Switzerland 2015   173
J.C. Westland, *Structural Equation Models*, Studies in Systems,
Decision and Control 22, DOI 10.1007/978-3-319-16507-3

Printed in the United States
By Bookmasters